I0187805

Canning and Preserving

What You Need to Know to Can Vegetables, Fruit, Meat, Poultry, Fish, Jellies, and Jam. Along with a Guide on Fermenting, Dehydrating, Pickling, and Freezing for Beginners

© Copyright 2021

The content contained within this book may not be reproduced, duplicated, or transmitted without direct written permission from the author or the publisher.

Under no circumstances will any blame or legal responsibility be held against the publisher or author for any damages, reparation, or monetary loss due to the information contained within this book, either directly or indirectly.

Legal Notice:

This book is copyright protected. It is only for personal use. You cannot amend, distribute, sell, use, quote, or paraphrase any part or the content within this book, without the consent of the author or publisher.

Disclaimer Notice:

Please note the information contained within this document is for educational and entertainment purposes only. All effort has been executed to present accurate, up-to-date, reliable, complete information. No warranties of any kind are declared or implied. Readers acknowledge that the author is not engaging in the rendering of legal, financial, medical, or professional advice. The content within this book has been derived from various sources. Please consult a licensed professional before attempting any techniques outlined in this book.

By reading this document, the reader agrees that under no circumstances is the author responsible for any losses, direct or indirect, that are incurred as a result of the use of the information contained within this document, including, but not limited to, errors, omissions, or inaccuracies.

Contents

Introduction

This book explores the various methods of canning and preserving food. It explains how you can pick, ferment, freeze, pickle, can, and dehydrate various foods. These include fruits, vegetables, meat, poultry, seafood, jellies, and jams. Many fruits and vegetables are seasonal, meaning they are not available or are less abundant in some seasons than in others. Many types of food spoil quickly at room temperature, in the presence of oxygen. By preserving food, you can keep it fresh for a longer period without it being contaminated by chemicals or harmful microorganisms. If implemented carefully and correctly, the food will retain its nutritional value, color, texture, consistency, and flavor. That is why thousands of years ago, humans started adopting methods to preserve food to safely consume it later. Besides delaying spoilage, making all types of food available throughout the year, and diversifying your diet, preserving partially processed food can save you a lot of time and energy when preparing it later.

This book is suitable for beginners and experts alike. It discusses and explains the purpose of each preserving method, how they work, and the benefits they offer. In this book, you will find hands-on and step-by-step instructions on how to prepare your food with different preservation techniques and how to apply them. The information is thorough, specific, and straight to the point, which will allow you to put it into practice confidently. The book provides advice on the type of equipment, tools, and containers that you should use. It also offers tips and tricks on selecting the best produce and overcoming potential issues along the way. You will find several simple recipes and clear directions that will help you to perfect the food preservation process.

Additionally, the book will teach you how to correctly use the food after it has been preserved to make sure it maintains its quality. The information in this book is up-to-date and compatible with modern-day technology. It is also highly accessible and applicable—no over-the-top equipment or procedures are required. The information is divided into clear, concise sections to help you easily comb through and find the details you're searching for. This way, you can use it as a guide or manual as you get to work. You will find numerous examples to ensure you carry out procedures successfully. The book uses common and understandable language, and any jargon or specific terms are clarified and simplified.

This book is meant to be fun and easy to read. Not only does it provide extensive knowledge and helpful guidelines, but it's also filled with interesting facts and information. You will gain general knowledge, find generic and diverse tips to help you in the kitchen, and, of course, you will master the art of canning and preservation. This book aims to be a firsthand, attainable, and reliable handbook. It presents accurate and real information. The procedures are neither oversimplified nor exaggerated in terms of functionality, practicality, nor implementation. This way, you will know exactly what to expect before you jump right in. The book is designed to help you get the most out of your food and decrease food wastage. It intends to help you overcome everyday issues and find solutions to the most common problems you may encounter in the kitchen. It's a friendly guidebook for anyone looking to make their time in the kitchen less of a headache.

Chapter 1: Introduction to Canning and Preserving

It's funny how we always crave our favorite fruits and veggies during their off-season. Not being able to satisfy those cravings can be quite frustrating. But it doesn't have to be that way. Canning and preserving your favorite food can give you a way out. Canned and preserved food can taste just as good and be almost as fresh as the day you first preserved them. The best thing about this is that the process is quite straightforward, and many things can be canned and preserved—jams and jelly, fresh fruit and vegetables, meat and pickles. And all this can be done without adding any artificial preservatives.

Preserving food is a process that has been around for centuries. Modern technology has been a boon, making life very simple for us. While we can simply head out to the supermarket to pick out what we need, our ancestors had to preserve food to survive the winter. Historical evidence shows that people in the Middle East would preserve food by drying it in the sun as far back as 12,000 BC. While that process took a lot of time, it saved people from starving in the winter. The practice of drying food, especially fruit, was also common practice in ancient Rome. The Romans were very fond of dried food. In regions where there wasn't enough sunlight available, they built "still houses," making it feasible. Different approaches were used in different regions based on the means available to their inhabitants.

While preserving and canning food may seem to be one and the same, they are actually two very different processes. Preservation of food has been done for thousands of years. However, the process of canning food was developed quite recently in comparison. Canning food dates back to the 1800s and was invented by French confectioner Nicolas Appert in 1809. This procedure was invented in response to a need to preserve food for the Army and the Navy.

Canning and preserving food offers numerous benefits. While preserved food can easily be purchased from the market, commercial products neither offer the same quality nor the satisfaction of preserving food on your own. In addition, they usually are more expensive and often contain artificial preservatives. Whether you grow your own food or purchase food in bulk, food preservation can help you to stock up, save money, and enjoy your favorite foods all year long.

How is Canning Different from Preserving?

Canning and preserving food have gained widespread popularity, with more and more people appreciating the benefits. While the two terms are often used interchangeably, they are quite different from each other. Not only are the processes different, but they also offer different results. While preserved food may last a few weeks in a refrigerator, canned food can last a lot longer.

Preserving Food

Preserving food is a method of prolonging its life by destroying any active bacteria and inhibiting bacterial growth. This is done by treating the target food with heat, acid, salt, sugar syrup, or, in some cases, even smoke. The medium used to preserve food naturally depends on the kind of food being preserved. Sugar syrup is often used to preserve foods like fruits, jams, and jellies. Whereas vinegar and oil are used to preserve pickled vegetables. This allows for extending their shelf life by a few weeks and, in some cases, even a few months. However, one may still need to store them in a refrigerator. Even herbs like basil leaves, bay leaves, oregano, curry leaves, and coriander can be preserved by sun-drying or simply drying them in a microwave.

Drying fruits and vegetables is also considered preservation, where the objective is to dehydrate them. The removal of water helps inhibit the growth of bacteria, yeast, and mold. This is probably the oldest method of food preservation, one that has been practiced for many centuries. Traditionally this was done by leaving food out in the sun, and sometimes even by air drying, smoking, or wind drying. This process takes quite a lot of time. Modern technology like electric food dehydrators, microwaves, and freeze-drying have made the process much faster and simpler.

Canning Food

Methods of preserving food had already existed well into the 1700s, but none of those forms were ideal for use in the military. There was a need to develop faster, more reliable methods to make both transportation and storage easy. It was towards the end of the 1700s when a need for better ways of preserving food arose. Napoleon Bonaparte needed a way to keep his armies well-fed and kick-started the search for a better way of preserving food. But the solution only came into existence in the early 1800s with the invention of the food canning process. The process of canning food came into existence when Frenchman Nicolas Appert responded to a call from the French government to develop a solution to help support the army. Over time, canning became extremely popular and is now used for preserving fruits, vegetables, soups, gravy, sauces, and meat.

The canning process involves sealing a can or jar with a combination of what needs to be preserved and a liquid, which is water. This process aims to kill any microorganisms present in the food and subsequently seals the food from the external environment to shield it from any further growth. These are the very same microorganisms that cause the food to spoil, and by killing them, the shelf life of food can be extended significantly. This is done by first sealing the contents in a container and subjecting it to heat and pressure. When the jar is subsequently cooled, the air inside is cooled and compressed, sealing the contents completely from the external environment. This protects the food from microorganisms and any further contamination. Compared to preserving food by storing it in a solution like vinegar, or sugar syrup, canning offers a higher shelf life. It does not require the canned food to be preserved in a refrigerator.

Benefits of Canning and Preserving

Canning and preserving used to be common practice. But over time, the usage of these techniques decreased significantly. The fact that you have access to almost every kind of fruit and vegetable, even when they are off-season, has made these processes redundant. However, there are numerous benefits to canning and preserving your favorite fruits and veggies at home.

Prevents Decay

Canning food helps to prevent decay. The canning process introduces various chemical changes like moisture, acidity, salinity level, and pH level of the food, killing any active microorganisms and preventing further growth. Preserving food can extend its shelf life for weeks, while canning can help prevent long-term decay, increasing the shelf life for years.

Economically Feasible

Buying food can be somewhat expensive. Growing it yourself can help you to save money. Unfortunately, fresh vegetables and fruits spoil quickly if you do not take measures to slow that process down. Canning and preserving food allows you to enjoy your homegrown produce for a long time, helping you to cut down your expenses. Not only is homegrown food cheaper, but food that is preserved and canned at home is also very economical compared to frozen foods available for purchase.

Taste and Flavor

On top of being more economical, canning and preserving food at home retains a higher quality, which improves the taste. In addition, pickling food can introduce a distinct flavor that you can tweak to your preference.

Healthier Than Commercial Products

Consuming canned food may not be as healthy as consuming fresh food, but it is better than preserved food purchased from commercial establishments. When you carry out the canning and preservation process at home, you can be certain of the ingredients used and attest to their quality. This is not the case with commercially canned products, and, by purchasing them, you can often end up with a lower than expected quality. Food canned and preserved at home will always be as healthy, if not healthier, than what you can purchase from retail stores.

Environmentally Friendly

Canning and preserving food can help to reduce your carbon footprint. This is because it can help you preserve excess food that would otherwise go to waste. By growing your own vegetables, you can also ensure there are no artificial preservatives and pesticides in your food. In addition, you also help to reduce the impact on the environment from large-scale commercial farming practices, and transporting vegetables from farmers to distributors, distributors to retailers, and retailers to your home.

Common Approaches for Canning Food

There are various approaches to canning and preserving food, each having their own upsides and downsides.

When it comes to canning, three methods are commonly used. These methods have been scientifically tested and proven to keep food safe and consumable for a long time. While the process of these three approaches may be different, they all follow the same principle.

1. Using a Boiling Water Canner

The boiling water canner is an ideal solution for the canning of food that has high acidity. The food is first placed in a jar which is then placed in boiling water. The heat from the boiling water helps kill bacteria and other microorganisms. This approach is generally used for canning fruits, jams, jellies, tomatoes, salsas, chutneys, and pickles. However, the same water canner may not be suitable for all types of food, and some may be specifically designed for jams and jellies.

2. Using a Pressure Canner

A pressure canner is generally used with vegetables and fruits that have low acidity. This is because it heats the food to a higher temperature and is more effective at killing microorganisms, helping preserve the food for a long duration despite a low acid concentration. Thus, a pressure canner is the perfect tool for potatoes, beetroots, green beans, corn, carrots, pumpkin, and meat. Two different types of pressure canners are generally used—a dial gauge pressure canner and a weighted pressure canner. In the dial gauge pressure canner, you can directly read the pressure from the pressure gauge and subsequently adjust the pressure by varying the heat. In the weighted pressure canner, the weighted gauge begins to hiss when a preset pressure is reached, which can then be used to adjust the flame and pressure as required. While both devices provide you feedback regarding the pressure in the vessel, a pressure gauge-based canner offers better control.

3. Using a Steam Canner

Unlike pressure canners, steam canners are used for fruits and vegetables that have a high acid concentration. The steam in an atmospheric steam canner can reach the temperature of boiling water and is just as safe for food canning.

It is important to note that it is always better to invest in specialized equipment for canning instead of rudimentary techniques such as heating in an oven or sealing hot food in a jar. These can be dangerous and ineffective.

Common Approaches for Preserving Food

Just like canning, there are different ways of preserving food items. Many of these processes can be easily performed by using equipment available at home. Here are some of the commonly used methods for preserving food.

1. Freezing

Freezing food you want to preserve is one of the most straightforward and simple ways of increasing its life span. However, there is more to it than simply placing it in your freezer. Before you decide to freeze your food, it has to be prepared. When it comes to fruits, they can either be frozen directly or with a sugar solution.

Freezing them with sugar can help to prevent discoloration and keep them fresher. Vegetables, on the other hand, must be cooked and blanched before being frozen. This involves heating them and then immediately immersing them in cold water. Doing this helps keep the vegetables fresher and prevents their aging. Finally, once the needed preparation is done, the food can be frozen, ideally in vacuum-sealed bags, as those can help prevent ice developing.

2. Drying

Drying foods essentially refers to dehydrating them. This can be done through various methods that include sun drying, wind drying, and using an electric food dryer. You can also use your microwave in some cases. For example, basil leaves can easily be dried and stored by placing them between paper towels and microwaving them in short bursts of a few seconds each. But you must be cautious because a microwave is not designed for this, and leaving it on for too long can burn the leaves, causing them to lose their flavor and aroma.

3. Fermenting

Fermentation of food has been done in a lot of cultures. Your favorite wine is made out of fermented grapes, and your favorite beer is made from wheat or sometimes even bread. But fermenting food doesn't just yield alcohol—it is also used to preserve food. For example, cabbage is fermented to make kimchi. Fermenting food also results in the formation of numerous vitamins and minerals. It makes certain foods more nutritious than their unfermented counterparts and makes them more desirable to some people.

4. Pickling

Pickling is the process of preserving food in vinegar, sugar, and other acids, which preserves it and gives it a distinct taste. It is commonly performed worldwide, and the process often includes using different spices to yield different flavors.

There are many different ways of canning and preserving different fruits and vegetables, each offering their own set of benefits. While canning offers long-term storage, preserving food not only helps extend its shelf life but, at times, also makes it more nutritious and desirable. Both techniques are quite easy to execute, but there are a few things you should keep in mind as they are crucial in helping your food retain its freshness and taste. You also need to use the right equipment to ensure the best results. From here on out, you will be introduced to various techniques and tricks that will help you to ensure your food lasts and tastes fresh. Canning and preserving food will not only help you to save money but will also help you to lead a more sustainable life by reducing your carbon footprint and reducing food wastage.

Chapter 2: Canning and Preserving Supplies

Now that you've had a small introduction to canning and preserving, it's time to get you acquainted with the supplies you will need. These are divided into four sections. First, the ingredients you'll be using regularly. Second, the appliances that will help you cook, can, and puree. Third, the necessary and extra tools and utensils you'll be working with. And fourth, everything storage-related, from jars to bubble poppers. All of that, along with a little surprise for those with a mind for business.

Ingredients

Pectin

Pectin is an ingredient common in all canning recipes. It is used to thicken fruit juice, leaving you with jam as your end product. While it is a natural component of any fruit, some fruits contain very little amounts, like cherries, while others contain large amounts, like apples. That's why you will often need to add powdered pectin to make any kind of jam. During the cooking process, it is released/added to the juices. Once in the juices, you should add the proper concentrations of lemon juice and sweetener to activate its gelling properties.

Bottled Lemon Juice

That's right, bottled lemon juice. For pectin to become gel-like, it needs a specific acidic concentration. When it comes to fresh lemons, you can never accurately tell how acidic they are, as it heavily depends on many variables related to the lemons' growing conditions. Bottled lemon juice provides a perfectly regulated alternative with a consistent pH (acidic concentration). That's why you'll find that most recipes call for bottled lemon juice, and that's why you should stick with it for better, more consistent results.

Sweetener

Who doesn't like a little sugar in their bowl? Sweeteners, whether sugar or honey, play an important part in the canning process for many reasons. In sweet preserves, the added sweetness amplifies the flavor of the fruits and counteracts the taste of vinegar. In savory preserves and pickles, it complements the acidity and adds a certain depth to the flavor. Sugar also contributes to producing a tastier, brighter-looking jam, jelly, or preserve, with much better consistency. That said, it is completely safe to cook without sugar, but make sure you use no-sugar pectin.

Vinegar

Adding white vinegar or apple cider vinegar is essential when it comes to pickling. However, depending on your use, you will either want to use high-quality vinegar or simply get the cheapest option out there. If you'll be using vinegar for pickling or adding a specific taste to a preserve, use the good vinegar and make sure it contains at least 5% acetic acid. On the other hand, if you'll only be using the vinegar to keep your jars from clouding during packing, the quality of vinegar you use won't matter (it won't come into contact with the fruits/vegetables).

Spices

If you will be playing with fruits and vegetables, don't forget to experiment with different flavors. Spices go a long way when matched with the right fruit/vegetable. The only thing you need to remember is to add the spices in their whole form (seeds, sticks, leaves) without grinding them. The flavor spreads more evenly when the spices are added whole, and even if the flavor isn't as powerful, it will be consistent. You can simply adjust quantity as desired.

Pickling Salt

What separates this type of salt from others is that it has no added chemicals, like iodine or other agents. As opposed to table salt, pickling salt doesn't affect the color of pickling water or the vegetables themselves. While other salt varieties won't hurt, as long as the salt you're using has added chemicals, your liquid won't be as clear.

Appliances

Water Bath Canner

What gives these canners an edge are two things: their large size and the fact that they come with a rack. The large size allows for canning the larger pint jars—several jars at a time, too. Meanwhile, the presence of a canning rack means you won't have to go and buy a separate one to protect your jars while canning. You should, however, keep in mind that this canner is the most suitable method for canning fruits and vegetables that are high in acidity. When it comes to low-acid produce, it's better to use a pressure canner.

Pressure Canner

Comparisons have often been made regarding the differences between pressure canners and water bath canners. The core difference is that pressure canners allow you to cook at very high temperatures by trapping steam inside and pushing water beyond its boiling point. The process is fast and efficient at deactivating all potentially harmful microorganisms. Pressure canners are the best for low-acid fruits and veggies because the extremely high temperature guarantees the death of certain bacteria known for producing the botulism toxin. In acid-rich fruits, the acid prevents bacteria growth and, in turn, toxin formation.

StockPot

Stockpots are the most versatile tool you'll find. Not only are they extremely useful during the cooking process, but if you don't have a canner and are not fully invested in buying one, stock pots are a great alternative. Granted, they are a lot smaller, but large stock pots can help you fill out a jar or two at a time. It's not too shabby for a substitute. Although, you will have to improvise when it comes to canning racks. Fold a kitchen towel and place it at the bottom of the pot. That way, your jars won't crack under the heat.

Slow Cooker

The slow cooker is the extreme opposite of a pressure cooker. It's perfect for turning fruit into complete mush. So, if you like fruit butter, salsas, and sauces, you'll love a slow cooker. More importantly, it's the best appliance to use if you don't appreciate the steam and the heat that a pressure cooker produces. Imagine cooking in a cool kitchen. That's the dream, right? Slow cookers are the perfect appliance for keeping your butter/preserve warm as you can it. Just set them to low and let them do their job.

Food Mill

Think of a food mill as a manual food processor. But it does a better and finer job when it comes to straining, grinding, and pureeing. If you aren't a fan of seeds, a mill is one thing you'll need to separate seeds from fruit chunks. When it comes to fruit butter, too, a food mill can help you puree the larger fruit parts and strain out the skin. It's especially helpful when it comes to making apple butter, jellies, and tomato-based sauces.

Utensils and Accessories

Colander

If you're planning on working with a large batch of fruits or vegetables, then you have a lot of washing, peeling, and rinsing to look forward to. Having a large enough colander will save you a lot of time and effort during those initial processes. Even deeper in the canning process, as you drain the hot fruit from its juices, you're still going to need a colander, more specifically, a stainless steel one. While plastic is cheaper, it can melt when exposed to high temperatures.

Jelly Bag

If one of the things you have your heart set on making is jelly, do yourself a favor and buy this type of strainer. The fine pores strain the liquid mixture, separating the solid components from the fruit juice, which can then be boiled with pectin, sugar, and lemon juice to make the jelly. You can also use a mesh strainer to achieve similar results, but it will take a little longer.

Wooden Spoon

Wizards wield staffs, and canners wield long wooden spoons. Because you're going to be doing a lot of stirring, you need a wooden spoon with a comfortable grip. You may have heard that wooden spoons retain bacteria. It's true; they do. After washing your spoon, don't use a cloth to dry the wooden spoon, but let it air dry to avoid any contamination from the cloth.

Knives

When it comes to peeling and slicing fruits, any knife will do, but some knives will make the job a lot easier than others. If you're not looking for accuracy, a chef's knife will do. For a more demanding task, like slicing or peeling small fruits, you want a paring knife. It is smaller, shorter, and lighter, making it easier to handle.

Peeler

Rather than using a knife to do the job, try using a hand-held or table-top peeler. You can also find specially designed peelers/corers that can help you easily de-core an apple. Unless you are an old-school chef who operates faster with a knife, or someone who appreciates a challenge, save yourself the hassle and invest in a peeler.

Towels

You are going to need many of them. Canning is a messy job, regardless of how neat you are or how organized you try to be.

Ladles

As mentioned before, plastic melts when exposed to high temperatures for a long time. A stainless-steel ladle is definitely the better option for you.

Storing

Jars

This is where your artistic side will get to shine. There are a variety of jars with different designs that come in different sizes. While the designs vary from one manufacturer to another, the sizes are pretty much standard, ranging from as little as 4 oz. (quarter pint) to as large as 128 oz. (gallon). Here are the most common sizes and their uses:

- 4 oz. jars: used for gifting preserves and storing baby purees.

- 8 and 12 oz. jars: used for storing jams, jellies, and preserves.

- 16 oz. (pint) jars: ideal for salsas and sauces.

- 32 oz. (quart) jars: ideal for pickled fruits and vegetables.

As for jar mouths, there are only two sizes: wide mouth and regular mouth. Wide mouths are perfect for storing pickles and bulkier canned products. If you intend to pickle whole or halved cucumbers, you shouldn't store them in a regular-mouthed jar for practicality's sake. Meanwhile, regular mouth jars will work just fine if you're storing a jam or preserve.

Lids and Rims

Every jar comes with its own lid and rim, but it doesn't mean you need to buy a jar to get a lid and rim. Each jar you reuse saves a lot of money and helps the environment, so, rather than buying entirely new jars, just buy a pack of rims and lids, mostly lids, to match with the jars you reuse. While the rim won't need to be replaced unless it starts rusting, the lid will have to be changed with each use. Lids are specially designed and lined with a sealing compound that seals a jar only once. Of course, you can opt for reusable lids, but they are a bit more expensive. It all depends on whether you are selling/giving away your jars or strictly keeping them for yourself.

Funnels

This nifty little gadget will save you a lot of trouble when it comes to transferring your preserves from pot to jar. When buying a funnel, opt for a wide-mouthed canning funnel. Most regular funnels have a wide mouth and a narrow neck. While you need a wide mouth to prevent spillage, fruits, veggies, and viscous liquids will get stuck in a narrow neck. On the other hand, a canning funnel has a wide neck that fits inside regular and wide mouth jars, making it easy for chunky liquids to flow through. You'll realize a canning funnel's true value once you start ladling your first batch.

Canning Rack

Having a canning rack for your jars is essential to carry out a proper canning process. To heat your jars (jar processing), you'll have to place them on a metal rack inside a water bath canner or a regular pot. If you place your jars directly on the bottom of the pot, you'll risk the jars breaking as glass doesn't react well to direct, excessive heat. In other words, a canning rack ensures the safe distribution of heat. Since the accessory is simply an elevated metal platform with a basic purpose, you are more than free to improvise. Canning rack substitutes include cooking racks, towels, and steaming racks.

Bubble Popper

After you fill up your jars with jam, you'll notice a few air bubbles. It's important to pop these bubbles to avoid upsetting the ratio between air and occupied space. When the jars are heated a second time, for vacuum sealing, the heat is supposed to apply pressure to the air molecules so that they escape the jar, sealing it. Trapped air bubbles interrupt this process, which is why you'll often notice that most recipes insist on bubbling. A bubble popper is a plastic tool that does the job perfectly. You could also use a wooden chopstick or a thin plastic spatula.

Budgeting

As you can see, canning is not cheap. This shouldn't be a problem if you are only looking for a hobby—you'll only need to know how to spread out your budget to get the necessary appliances. However, if your aim is to turn this into a profitable business, you will need to do some calculations before investing. It's all about expense projection.

Expenses are divided into two types: fixed and variable expenses.

- Fixed expenses are those which will cost you the same amount, whether you're making large quantities or not. These include the prices of the appliances used for storing, like freezers and fridges. Even the cost of electricity required to operate a freezer, as well as the costs of appliance repair, should be included in your fixed costs equation.

• Variable expenses are those which vary according to the quantity you're producing. Let's say you make a 9-pint batch of peach jam. You'll have to buy roughly eleven pounds of peaches. You'll also have to buy enough jars, lids, and rims. Your utility bill will also increase according to how much gas/electricity you use. Meaning, the more you produce, the more your variable expenses will increase.

When figuring out a budget and the pricing of your product, you'll need to calculate an average of your expected fixed and variable expenses.

Fixed Costs	
Item	*Total / Batch*
Equipment (freezer, fridge, etc.) (cost/expected lifespan)	
Fixed equipment—operation costs	
Yearly equipment repair budget	
Total Cost per Jar (total fixed cost/number of jars)	
Variable Costs	
Item	*Total / Batch*
Ingredients	
Variable equipment—operation costs	
Packaging (Jars, rims, lids, labels)	
Delivery (if applicable)	
Variable Cost per Jar (total variable cost/number of jars)	

The sum of the two tables should make for your expected starting cost. In research done by the University of Florida in its *Food and Resource Economics Journal*, a 2018 budgeting study found that an 8-half-pint-jar batch of strawberry jam costs $26.46 per batch. It also took a price of $3.31 per jar only to break even (no profit or loss). This will likely vary with market pricing. Nevertheless, costs vary from one person to another depending on whether they can find ways to cut costs (buying in bulk, saving on operational expenses by cooking in large batches, and reusing jars).

Chapter 3: Picking the Best Ingredients

Now that you know about all the supplies that you'll be needing to start your canning and preserving adventure, it's time to move on to the most important element, the ingredients. The quality of your ingredients is key if you want great results. There's a common misconception that the produce used doesn't need to be fresh or of high quality if it's going to be preserved or fermented. This is completely wrong. Picking the right ingredients is more of a science-backed skill that you should learn more about before giving it a go. This chapter will give you lots of tips and tricks to start with until you gain some solid knowledge to identify the best ingredients at first glance. To make it easier for you to follow up, we'll classify ingredients into three categories—fruits and vegetables; meat, poultry, and fish; and finally, jams and jellies. Let's get started:

Fruits and Vegetables

Fruits and vegetables are the most common ingredients used for preserving as they are relatively easiest to handle. Though it can be a lengthy process, you'll find it very useful to grow your own produce and enjoy your summer veggies during the cold winter months.

Tomatoes, cucumbers, and squash are essentially summer vegetables. However, with the preservation methods that you'll learn throughout this book, you'll be able to enjoy them year-round. While most fruits and vegetables can be preserved, some give better results than others. As a general rule, high acidity makes for a better candidate for canning and preserving. This includes berries, apples, peaches, pears, plums, and nectarines. As for the vegetables, tomatoes, potatoes, mushrooms, and carrots top the list of the best-performing preserving ingredients. Regardless of your ingredient pick, you can follow some general guidelines to select the most suitable ones. Here are the top factors that you have to consider when choosing your fruits and vegetables:

- **Ripeness:** your produce should be fully ripe. Otherwise, you'll be disappointed that the taste and texture are off. If you usually buy your produce from local vendors, consider making a deal with them to set you aside some that will fare well with preserving.

- **Free of defects:** one bad piece of fruit or vegetable can invite harmful bacteria to fester and ruin your entire batch. So, be extra-cautious when deciding which ones to use. Choose produce that is not bruised or damaged, and which feels sturdy to your touch. You should also consider its skin and flesh color—it has to be bright and healthy, showing no defect signs.

- **Aromatic:** aroma is a very important element that sets apart good produce from bad produce. When you're at the market, put your sense of smell to work. A quality mango should fill the air around you with its intoxicatingly sweet and overpowering smell. The same goes for watermelon, honeydew, and peaches, just to name a few. In general, a more pronounced aroma most probably means higher quality produce.

- **Chemical-free:** this shouldn't be a problem if you're growing your produce yourself. However, organic, chemical-free fruits and vegetables can be quite expensive to buy at the market. But, they make much better ingredients for preservation. Besides, when you're looking to preserve produce, you're only making a one-time bulk purchase. So, it's not going to have a big impact on your bank account at the end of the month.

- **In-season:** this might be an obvious factor. However, it's worth reminding yourself to use in-season ingredients if you want the best results with canning or freezing or any other preservation method. If you're a pasta lover and can't imagine ever running out of fresh tomato sauce, you should preserve your tomatoes during summer. Tomatoes love sunshine and clear air, so they grow best during the summer months. However, given the high demand, you might want to make some special arrangements to get the quantity you need for preservation. Make sure to plan ahead and have all the preservation supplies and tools at hand to make the most of your juicy summer tomatoes.

Fruits and vegetables are very versatile when it comes to preservation. The method that you choose comes down to your preference. However, here are a few suggestions recommended by preservation fanatics.

If you want to give canning a go, we mentioned above that berries, apples, and pears make for excellent options due to their high acidity. Whole tomatoes remain the most common choice.

If you're more interested in fermenting, root vegetables like carrots, beetroots, turnips, and radishes, and of course, the super healthy fermented cabbages (sauerkraut) should all be on your list. They all withstand the fermentation process incredibly well and can fulfill your dietary probiotics requirements much better than any supplements. Fermented fruits are more of an acquired taste—not everybody seems to enjoy them. However, plums, peaches, and apricots are typical choices for fermentation as their color and shape stay intact. Fermented apricot leather is especially famous in the Arab world, where it is used to make one of the infamous Ramadan drinks, Kamar Eldin.

Dehydrating fruits and vegetables is one of the oldest preservation techniques. Dehydrated bananas, apples, strawberries, and mangoes are some of the healthiest snack options to pick up at the supermarket. Zucchini and sweet potato chips are also all the rave right now. However, they'll cost you some serious cash if you get used to treating yourself to these tasty crisps. By the end of this book, you'll be able to prepare them yourself for a fraction of the price.

Freezing is the universal option for produce. In fact, it's believed to be the healthiest among all the preservation options, since frozen foods retain around 97% of nutrition. The list of freezable fruits and vegetables is very long and includes strawberries, berries, bananas, kiwis, peas, corn, broccoli, cauliflower, and winter greens.

Meat, Poultry, and Fish

Many people are unaware of the fact that you can safely preserve meat at home. It's actually a great way of making sure you always have some canned beef on hand to prepare your favorite soup or noodle dish. Just as with fruits and vegetables, there are a few important factors to bear in mind when choosing which meats, poultry, or fish to preserve. Always make sure your meat is:

• **Fresh:** while you can preserve pre-frozen meat, it's better to choose meat that has just been butchered for the best result. Don't spare any time or effort scouring your neighborhood until you find the best local butcher. Ask to see all the necessary tags and licenses to ensure that your meat is top quality.

• **Low-Fat:** lean meats and poultry preserve better than fatty meats. The extra fat lingering around the canning jar gets in the way of proper sealing and increases the chances of your meat going bad. Prep your meat accordingly before you start preservation. Cut out any excess fat and get rid of bruises.

- **Boneless:** if you're canning beef, lamb, or pork, it's recommended that you get rid of the bones. However, for chicken, duck, and other kinds of poultry, there's no need for deboning.

Below, you can find some examples of the kinds of meat you can use and the most appropriate preservation method for each.

Venison Cubes or Strips

Having some pre-prepared jars of preserved venison cubes will save you a lot of time in the kitchen. Canning is the favored method for this specific meat cut. You can either go for a raw or hot pack. As the name suggests, if you use a raw pack, you won't need to precook your venison—simply adding some canning salt on top of your raw meat in a jar will get the job done. Remember not to add any kind of liquid to your raw meat as it releases its own juices with time.

If you're a fan of the caramelized meat flavor, canning with a hot pack will satisfy your taste buds. This method is done by giving the venison cubes a quick sear before placing them in the jar. You'll notice that the jar will fit more meat with this method as the heat makes the meat cubes shrink in size.

Ground Beef

You don't have much choice with ground beef—you'll have to use a hot pack. Use an electric meat grinder or a manual one if you want to make up for missing any workouts. Grind your beef chunks, then heat the meat to a nice brown color on the stove, optionally with chopped onions and garlic if you like the taste. Add the liquid of your choice (tomato juice, water, or broth) to your ground beef and bring the mixture to a boil. Once the liquid is reduced and the ground beef takes a nice brownish color, divide it into canning jars, adding some canning salt on top. Again, you can choose whether to add water or broth, then seal the jars and pop them into your preheated pressure canner.

Sausages

Sausages are perhaps the most commonly used meats for fermentation. Despite having a bad reputation amongst health fanatics, who usually complain about their high-sodium content, they're still wildly popular worldwide. In fact, fermented sausages were found to contain a significant amount of heme-rich protein. If you fancy this kind of meat, preparing it at home can be a good way for you to control the sodium content. After getting rid of the fat in your meat, you can spice it using your favorite herbs and spices, then add a bit of sugar. Sugar will help with the production of lactic acid that is essential for the fermentation process. You can then stuff your casings with the ground meat mixture, making sure to remove the air bubbles. Hang the sausages to air-dry somewhere with high humidity. After a few days, you should be able to enjoy your homemade fermented sausages.

Chicken

Chicken, duck, or turkey are all great choices for raw pack canning. Unlike hot pack canning, which tends to over dry the tender poultry meat, the raw pack keeps the meat moist and juicy. Keeping the bones in or taking them out is your choice. However, you have to remove all the skin.

For safety reasons, it's better to use chicken that has been dressed and chilled. When canning poultry, a rule of thumb is to always go with bigger birds, since they tend to be packed with more flavor that remains intact with preservation. From here onwards, the process is pretty much the same as the canning of venison. After cutting the chicken, you simply put it in the jar, add some canning salt, and hold off on the liquid. Along with every other kind of white meat, chicken also fares well with dehydration, thanks to its low-fat content.

Fish

If you fish for your own shrimp and tuna, you don't necessarily need to freeze or can them as you may already be accustomed to. Dehydration is an excellent and relatively easy preservation method to enjoy shrimp, tuna, and even imitation crab. To kill off any harmful bacteria, you have to make sure you dry your fish at 145 degrees Fahrenheit. You should also remember to remove any excess fat to avoid rancidity.

Clams

Many people love clams but hardly have access to them. If you're a fan of this seafood delicacy, you can preserve your own clams at home. Canning clams is a simple process, but it does include several steps. First, you should keep the clams alive until you're ready to start canning. After washing and scrubbing the shells really well, steam them for a few minutes until they pop open. Next, scrape off the meat and set the clam juice aside. Give the meat a good wash and boil it in hot water with some lemon juice and half a teaspoon of citric acid. After boiling for two minutes, strain the water and add the clam meat into the canning jars. Then add the clam juice that you saved earlier on top of the meat, seal the jar, and you're all done.

Jams and Jellies

To make delicious jams and jellies, you have to be picky with the ingredients you choose. Some fruits may seem like they will make for a good base to prepare your jams. However, small details like their acidity and water content affect the final result. To help you make the right choice while grocery shopping, you should use the tips below:

• Choose fruits that are firm and free of any defects. Ripe fruits have good amounts of pectin, which is responsible for the stickiness and gooiness of jams after they are set.

• Give the fruits a taste test. Whenever possible, you should always try to give the fruits a taste before buying them to ensure that the flavor is just right.

• Purchase the right quantity. There's really no point in going overboard with the quantity of fruit you buy on account of its freshness. Cooked fruit will not be good to eat for too long without being preserved. Aim to buy the exact amount of fruit that you'll be using in your jams and jellies.

• Choose in-season fruit to prepare good quality preserves.

• You can opt for frozen fruit to prepare jams. Whether homemade or store-bought, all you have to do is thaw your fruit ahead of time, so you'll be able to crush and use them for preservation.

If you want to try your hand at making jams and jellies, you'll be happy to know that there are many options to choose from.

Grapes, apples, berries, pears, and of course, strawberries are only a few of the fruits that you can use to make jams and jellies.

As you can imagine, choosing high-quality ingredients will make a world of difference in the final results. If you're a beginner, the whole thing may seem a little intimidating at first. However, if you use the tips mentioned above and use them as your personal guide when you are at the market, you'll be much more confident when you head back to your kitchen. The idea is to look for the best ingredients you can get your hands on. Don't think that there won't be much of a difference in the end, because as you well know, there will be. In the next chapter, you will go on a journey to dig deeper into what water-canning is all about. You'll understand everything about the water-canning process, how and when to use it, and all the necessary supplies that you'll be needing.

Chapter 4: Water Bath Canning

So, you've learned how to pick the best ingredients to make delectable jams and jellies. You've bought the supplies, and your set-up is perfect. Now it's time to get your pantry stocked with jars of your homemade creations that your whole family will enjoy. The first thing you need to master at this stage is water bath canning, which will help you get shelf-stable jams and pickles free of any preservatives. This chapter will focus on the water bath canning process and help you figure out the details step by step. While it may sound like a pretty daunting exercise for a first-timer, it's actually a lot of fun once you get the hang of it.

First Things First...

A quick reminder before you start this process: water bath canning is only truly effective with foods naturally high in acid, including most fruit preserves like jams, jellies, fruit canned in syrup, and most kinds of pickles. Of course, you should not venture into jam-making without making sure that you are relying on a recipe from a reliable source. Never attempt to undertake this process with non-acidic vegetables, soup, different kinds of stocks, meat, fish, or poultry. These require more complicated processes utilizing a pressure canner. Water bath canning does not apply to these kinds of food.

Equipment and Set-Up

With that warning out of the way, we can get to the more interesting stuff. Let's start with the equipment you'll need and the proper way to set up. A few things are probably already in your kitchen, which is great—you don't always need to invest in pricey equipment and can simply work with what you have. And, while there are pre-made canning kits available for sale, you don't have to splurge since you can easily rely on things already available in your home. This can definitely help you save money in the long run.

Simple Set-Up

You'll need a few basic things: a tall pot (the sort you'd use to boil lobster or prepare shellfish, for example) and a rack to fit inside the pot. Of course, you'll also need a whole bunch of canning jars with two-piece lids—the most popular kind is simple mason jars.

You will also require a ladle, a funnel to help with all your canning endeavors, a timer, some tongs, a clean spatula, and lots of clean towels. Although it's a less environmentally friendly way to carry out the process, you can use paper towels.

Additional Equipment

The list above is technically all you need to get started. However, other tools can help to streamline the process. Some of these items are a lid caddy to help you keep lids organized, and a magnetic lid wand, which can be a lifesaver. This handy tool helps you to remove the sterilized canning lids from the boiling water to prevent contamination, all while keeping your fingers away from hot water. This will keep you from having to suffer through a bunch of unnecessary burns when making your way quickly through the jam or jelly canning process. Also, you may need a canning rack with handles to help make things a bit easier, and a stovetop or electric kettle will also be useful when you need to add more hot water quickly.

Starting the Process

The first thing you need to do is place the rack at the bottom of the tall pot—the rack helps to keep all the jar bases at a safe distance from the bottom of the pan. Instead of burning or breaking the jars, the water can evaporate and escape around the jars, which prevents them from shaking and knocking against each other, causing them to break. Add enough water to cover the jars, at least one inch above the lids. Turn on the heat and bring the water to 140 degrees Fahrenheit. If you are hot packing, make sure the water is at least 180 degrees. Use a thermometer to get the necessary accuracy. You can begin this part of the process while the food to be canned is being prepared.

Filling the Jars

This sounds like such a straightforward step that it's almost laughable. However, filling the jars with preserves is the more detail-oriented part of the job, and it needs to be done carefully. First, you need to check the jars and rims to ensure they don't have any chips or imperfections after being inside the pot. Also, make sure to use new lids that haven't been used before. The rings can never be reused—they're really only good for one go.

A quick tip: look at the instruction manual for the jars and lids before using them. Some may require you to heat the jars lightly in a water bath while the lids are kept in a separate container of hot water. The ever-popular mason jars have recently changed their recommendations, stating that this step is no longer necessary when using them. But many other brands still require that this initial step is performed the old-school way, so check first.

All jars and lids need to be cleaned with hot, soapy water, rinsed, and properly dried right before filling them up. If this seems too cumbersome, you could also run the jars through a cycle in the dishwasher. This will ensure that they are cleaned with the hottest water possible, maintaining the "bath" element crucial to stabilizing jams and jellies.

After washing, you'll need to work fast so that your canning jars and lids remain at the required temperature. Quickly fill the canning jars, leaving an adequate amount of space at the top between the food and the rim of the jar. Usually, somewhere between a quarter of an inch and an inch is safe. Be sure to check your recipe for the exact details, since this can also vary according to the different recipes you may be using.

Topping Things Off

Once you've added the concoction, be sure to run a clean spatula along the interior of the jar to help release any trapped air bubbles. Then, make sure to wipe any trace of the food off the rims using a clean, moist towel. As they say, the devil is in the details, and wiping away any trace of the food off the rims will allow for much better contact between the lid and jar, which will give you the perfect seal.

Next, you'll want to place the round canning lids onto the jars. If you manage to get a magnetic lid wand, you'll find that this will be a huge help. It allows you to grab them one at a time quickly and efficiently. Basically, you'll want to screw them onto the jars as steadfastly as possible so that they are screwed tight. However, you also want to make sure they are not so tight that too much air escapes the jars, which may compromise the quality of your food.

Processing the Jars

Next, you'll have to boil the jars in the same way outlined above, this time with the jam or jelly inside. In the first step, it was done primarily to sanitize everything before adding the food. In this step, however, it's meant to further stabilize the product inside in a natural manner, without the addition of preservatives, unlike most store-bought products.

This time, you just need to load the jars onto the rack and lower it into the water bath slowly. If your rack doesn't have handles—a tool that comes in handy when adding your jars into already boiling water—then use tongs to place them. Be careful and make sure the jars are kept vertical so that the food doesn't come into contact with the rim of the jars. Again, there should be at least one inch of water above the top of the jars. If you don't have enough water, quickly boil some in a kettle and pour it into the pot. Be sure not to keep the jars too close—you will want to keep at least half an inch of space between each of them.

When you're done, turn up the heat and bring the water to a full, rolling boil, putting the lid on the pot. Start the timer according to the recipe you're following and keep a close eye on what comes next. Most canning recipes will ask that you boil a water bath for about ten minutes. Keep in mind that the processing time doesn't begin until after the jars are fully submerged, and the water has come to a full boil. If the water is boiling too much, causing the jars to shake and rattle, then reduce the heat till it boils at a gentler pace. But make sure that it is still boiling.

The Cool Down

When the processing time for your water bath canning has ended, turn off the heat, let the jars find their equilibrium, and settle for about five minutes. Set the timer again to help you keep track.

Then, remove the jars from their rack using the tongs, lifting them vertically, and be very careful not to tilt them. They're at a vulnerable stage, so be careful with them. Otherwise, the food will touch the lid, something you do not want at this stage since you're still trying to make sure everything will be shelf-stable, with a perfect seal.

Transfer the jars to a cooling rack, or simply place many towels on the counter to protect your surfaces from the hot glass. Make sure to keep a distance of at least one inch between each jar. The towels or cooling rack is very important because you don't want the temperature of the jars to drop suddenly. The temperature shock can cause them to break. Make sure that your jars are left without anyone shaking them or moving the contents around for at least twelve to fourteen hours until they've completely cooled. Any movement at this juncture will cause the lids to flex, which can break the seal, so keep them still. Once in a while, the lid will make a small sound. This is the seal settling with the new temperature and is generally a good sign.

Storage

Once you've let them cool, test the seal by gently lifting the jar by gripping the seal—you should be able to easily lift the jar from the lid. If the lid falls, then put it in the fridge and make sure to consume the food right away. Otherwise, you'll be leaving your work out to rot. If you haven't left the jars out for too long, you can get the contents reprocessed within 24 hours.

If all goes well with the seal, wash and wipe down the jars with a moist cloth and store them in a cool, dark place. They can be safe to eat for at least a year after you're done. Now your family can enjoy the fruits of your labor any time and taste the freshness of sweet strawberries even when they're not in season, or have the best kosher pickles without having to go to a deli. Preserves made at home have a distinct flavor that cannot be replicated elsewhere. Once you've managed the process of canning in water baths, you will be well on your way to having a perfect harvest all year round.

Chapter 5: How to Pickle Any Fruit and Veggie

Pickling is a food preservation technique that dates back to 2400 B.C., according to the New York Food Museum. Four-thousand years ago, the Mesopotamians couldn't preserve their foods for long periods, given that freezing technology did not yet exist. They were often at the mercy of their yield. You could say that the frustration of throwing away food one month and starving to death in the next pushed them to innovate. They started putting their foods in containers with vinegar. Some also started using brine as a preservation liquid, and it worked.

These extremely acidic, salty solutions worked because they created an inhospitable environment for bacteria and fungi to grow. Microorganisms can only survive in specific temperatures, acidity, and salinity levels. Any drastic change will deactivate them, if not kill them. That's why people boil milk to sterilize it, use a pressure canner to sterilize low-acidity fruits, and put cucumbers in brine/vinegar.

Over the years, what started as a necessity, became a way of life and a luxury and delicacy. People realized that pickling changed the texture and taste of a vegetable and fruit. Not just that, but they also discovered the full power of spices when it came to adding flavor. Put two and two together, and there you have it, the origins of pickling.

When Napoleon, needing to provide sustenance for his army, asked for a preservation method, Nicolas Appert stepped in. In 1809, he simply suggested removing air from the jars before sealing them. In other words, he suggested water bath canning to create a vacuum seal.

About 91 years later, in 1900, another food preservation method—fermentation—gained prominence. Compared to pickling, fermentation relied on the acid produced by good bacteria within the fermented foods, rather than adding vinegar or any other external source. As you can see, both methods are based on the same concept of creating an inhospitable environment for harmful bacteria, but the techniques can be very different.

This chapter is solely dedicated to the culinary art of pickling. So, get ready to learn the ins and outs of the pickling process. By the end of the chapter, you'll be able to pickle and can any type of food you want.

What You Will Need

There are two types of pickling: quick pickling and regular pickling (the pros' choice). The main difference between the two is in canning. The brine is made and added to the vegetables in a jar with quick pickling, then put in a fridge. The pickles only last for about a month. On the other hand, regular pickling uses a higher vinegar concentration and, more importantly, involves a water bath canning procedure. The procedure removes oxygen from the jar and vacuum seals the pickles. Combined with the vinegar, the water bath canning keeps the pickles safe to consume for up to a year (no refrigeration needed). Not to mention, regular pickles have a stronger, more multi-dimensional flavor.

Regardless of the process you choose to try out, you will need a few basic ingredients: white vinegar or apple cider vinegar—dealer's choice—pickling salt, sugar, whole spices, and water. It goes without saying that you also need the food that you will be pickling.

Both processes also have one instrument in common, and that's a saucepan for preparing the pickling brine. As for your jars, it entirely depends on what you want to pickle. Whole veggies/fruits, like cucumbers, require large jars (16 oz. or larger) with a wide mouth, so you don't struggle when packaging them or taking them out. You can use regular mouth jars for smaller/chopped-up foods, like onions and carrots.

If you want to follow through with the longer canning process, you will need a few more things. As mentioned before, you need a water bath canner because it's where the process takes place. Second, you need a canning rack or a folded kitchen towel to place under the jars when they are heating up. Third, loads and loads of water, and a little more vinegar. Last but not least, a couple more towels for when the hot jars are out of the canner.

Quick pickling should seem much more appealing right about now, especially after knowing all the work required for regular pickling. Quick pickling does save a lot of time and effort, but before you opt for the shortcut, consider the scenic route. It's hard work, but you'll be much more satisfied with the end product.

Preparing the Brine

Brine is the pickling liquid. It's the vinegar, salt, sugar, and water combination that makes or breaks a pickle. Suffice to say, you have got to make a fine brine—no pressure. As important as they are, brines are simple, and the key is to get all the ratios right. When starting, you want to learn how to make a basic brine—not too sugary, not too acidic. As you start experimenting with different fruits and vegetables, you can tweak the ratios to accentuate the flavor of the food you are pickling.

Ingredients

- 1 cup of white or apple cider vinegar (at least 5% concentrated)

- 1 cup of water

- 1 Tbsp. of pickling salt (not table salt). You can also substitute pickling salt for kosher salt. The conversion ratio is 1:1.5 (pickling to kosher salt). In other words, exchange each tablespoon of pickling salt with 1.5 tablespoons of kosher salt.

- ½ cup of sugar

Pro tip: if you can't estimate how much brine you'll need, pack your jars with the fruits/vegetables, then fill them up with water. Pour the water back into a measuring cup, note the amount, and split it between vinegar and water.

Directions

1.   Put all of your ingredients in a saucepan over high heat. Add your herbs and spices if you want to add any.

2.   Bring the mixture to a boil while stirring the sugar and salt until they've dissolved completely.

3.   Turn off the heat and divide the mixture across your jars (filled with veggies). Make sure you leave about ½ an inch of space between the top of the jar's contents and the inside of the lid.

4.   Stop here if you are quick pickling. Wait for the jars to cool down before you put them in your fridge. Eat after a day or two to get a stronger flavor.

5.   If you are regular pickling, you'll need to de-bubble your jars because the air bubbles compromise the vacuum seal. Tap the jars lightly on your counter, then stick a bubble popper (a wooden chopstick/utensil will do, too) around the sides of the jar to remove any air pockets. You might need to top off your veggies with more brine after the air escapes.

6.   Wipe the jars clean for a better seal and start the canning process.

This brine mixture will allow you to pickle anything you want without having to change the ratios. Still, you're free to adjust them to your personal taste.

Note: While quick pickling ratios are flexible, canning ratios are not as flexible. Keep in mind that each canning recipe and brine ratio is designed to maintain a certain level of acidity to preserve food safely from botulism-producing bacteria. When tweaking your recipes, make sure you maintain a ratio of 1:1 (vinegar to water). You can always increase the amount of vinegar or decrease the water, but don't dilute the vinegar. As for the other ingredients, you're free to improvise however you want.

The Possibilities Are Endless

What can you pickle? Aside from human body parts—for legal purposes—you can pickle anything. Vegetables: from asparagus to zucchini. Fruits: from blueberries to grapes to watermelons. Amsterdam is known for its pickled herrings, Louisiana for its pickled pork, and Germany for its pickled eggs, which have made their way to

Britain and the United States with the Pennsylvania Dutch. In other words, the possibilities are endless and overwhelming.

Best Veggies for Pickling

• **Cucumbers (technically a fruit):** the cucumber is so famously pickled that it has become the face of the word "pickles." There are several ways to pickle a cucumber. You can cut the bitter tips and pickle them whole or slice them vertically in halves or quarters. If you're feeling a little cheeky, you can go for pickle chips. Use a crinkle knife and cut your cucumbers in circles.

• **Red Onions:** pickled onions are beautiful. They are a little sweet, a little sour, and can make a regular burger taste like a million bucks. Just thinly slice your onions into semi-circles and consider increasing the sugar in your brine just a little.

• **Asparagus:** definitely an interesting choice of vegetable, and you've got to blanch them first to soften them. Drop the stems in salted boiling water, wait for about three minutes, then drop them in iced water for at least five minutes. Make sure your jar is tall enough, though.

Best Fruits for Pickling

• **Peaches:** all picnic fruits bow before pickled peaches. Boil one cup of vinegar with two cups of sugar (no need for salt or water), then add the peeled peaches into the mix. Cook until the peaches are soft, and there you have it. For a special touch, boil 3-5 cinnamon sticks with the vinegar-sugar-peach mixture.

• **Apples:** what goes well with a scoop of ice cream? Pickled apples. Use a basic brine and go easy on the salt (1 tsp. instead of 1 tbsp.). Boil your ingredients, then bring them to a simmer before you add the apple wedges. A few cinnamon sticks will go along nicely with the apples, too.

- **Blueberries:** the perfect companion to all things cheese, from cheesecakes to brie sandwiches and salads. Add the vinegar and spices first, then pop the blueberries into the saucepan once you bring the liquid down to a simmer. Let cook for five minutes until the berries start swelling, and no matter what, don't stir.

Shake the pot to move the berries around, but don't stir, or you will damage the fragile berries. Turn off the heat and let cool for a minimum of eight hours before you drain the berries and ladle them into jars. Next, put the drained liquid back into the saucepan, add the sugar, and boil for three to four minutes for a syrupy brine. Finally, pour the brine over the berries and seal the jars.

Interesting Combinations

Who said you couldn't pickle more than one veggie/fruit together? There are endless combinations. It's all about finding the things that go together. Sweet fruits go well with spicy veggies, like peaches and habanero peppers. Another complementary power couple is apples and ginger slices. If you're not much of a sweet tooth, you can pickle avocados with chili; the perfect mix between buttery and spicy. Once you run out of complementary pairs, try putting together fruit/veggie groups that share a similar feel, like peaches, plums, and grapes, or carrots and onions.

As mentioned before, the options are endless. It's all up to your imagination. As long as vinegar occupies at least 50% of your brine, you can put anything in that jar. Go ahead and try out these fruits and veggies, or go rogue from the start. Who knows, you might stumble upon a delicious pickling recipe. It's time to get creative.

For an Extra Kick

Every master pickler knows their way around a spice cabinet. Dill, hot pepper flakes and black peppercorns can turn a two-dimensional jar of cucumber pickles into a taste fest. Pickled plums coupled with cinnamon sticks, star anise, and cloves can spread warmth through your body with every bite. That's the power of herbs and spices.

Spices	Herbs
Hot Pepper Flakes	Rosemary
Cinnamon Sticks	Dill
Star Anise	Mint
Mustard Seeds	Lavender
Fennel Seeds	Basil
Coriander Seeds	Thyme
Bay Leaves	Cilantro
Vanilla Bean	Marjoram
Ginger Root	Parsley

The most important thing to remember when dealing with herbs and spices is to trust your senses. Take your time testing out various spice and herb combinations. Play around with the different flavors and intensities as much as you want, but remember to always use whole herbs and spices. Ground spices always result in a cloudy, unappealing brine. Other than that, there are absolutely no wrong answers.

What to Do with a Half-Full Jar

Once you've eaten your way through your first jar of pickles, you're going to face a tough decision. How do you bring yourself to throw out the tasty brine you worked so hard to make? The good news is, you don't have to. If you're going to learn how to make the brine, it's only fair to learn how to make the most out of it, and there are, figuratively, a million ways to do so.

More Pickles

If you properly pickled the first batch of pickles, why not use the leftover brine to quick pickle a second batch? It will save you a ton of time, effort, and money.

Instant Salad Dressing

Have you ever wanted to put together a quick, refreshing salad to balance out a cheesy/creamy meal, but got turned off at the thought of preparing a dressing? Just pour a little brine over your selection of chopped veggies and enjoy.

Alcohol it Up

Who said you can't get drunk on pickle juice? Two parts vodka or gin (two shots/2 oz.), one-part pickle juice (1 oz.), some ice, and a pickle slice as garnish—and you have one pickle-tini recipe you'll want to try.

Pickle brine can also serve as a secret ingredient for the Bloody Mary recipe you've been trying to perfect for years. It's acidic and salty, with just a hint of sugar that can unlock a whole new range of flavor.

If you're not a fan of cocktails, chase your whiskey with a shot of pickle juice to rid yourself of the taste and the burn. The acid in pickle juice neutralizes the taste of alcohol.

The Morning After

After having all those cocktails, you should be expecting a hangover, but it shouldn't be a problem. Pickle brine helps with that, too. After your body is sucked dry of its water content and your whole electrolyte concentration is off-balance, you're going to need a dose of electrolytes. What's better than vegetable water, vinegar, and salt to replenish your electrolytes? However, before you dig in, cut the brine with more water or club soda. Too much acid can irritate your stomach lining.

With this information in your back pocket, nothing can stand in your way of becoming a master pickler. You only need practical experience, and there's only one way to get that, so bring out the jars and get pickling.

Chapter 6: How to Make Homemade Jam and Jelly

Besides pickling, making jam and jelly are the most popular ways to preserve fruit. If you manage to get your hands on some good quality ingredients, it's quite easy to prepare these enjoyable, sweet fruit spreads. You can store them for a couple of weeks without canning, but if you seal them with water bath canning, they can last till the next season.

This chapter focuses on teaching you how to make homemade jelly or jam from any fruit in a few easy steps. You might find the process a bit overwhelming at first. Still, nothing can outweigh the benefits of preparing these natural products yourself. They won't contain any artificial colors, flavors, or preservatives, after all. And honestly, despite all the artificial flavor enhancers in the store-bought spreads, homemade ones taste a lot better. Plus, it costs far less to make your own fruit spreads and store them in your pantry than to buy them from a store.

The Supplies You Will Need

To make your own jam or jelly, you will only need some fruit (fresh or frozen), sugar, water, acid, and pectin. The last ingredient is optional for some jams but necessary for making jelly. You will also need at least one saucepan large enough to cook the fruit, some sterilized glass jars with lids and screw bands, and a cheesecloth if you are making jelly. After you have all your supplies, arm yourself with patience, because cooking can take a long time. Especially if you plan on making a large batch of jam without using any pectin.

The Difference between Jam and Jelly

Although made with the same ingredients, jams and jellies have different consistencies. A jelly typically tends to be much smoother and firmer than a jam and has a clearer appearance. The reason for the difference lies in the process of making these two spreads. The finished jelly contains only fruit juice, without pulp, while the finished jam has the whole fruit inside. That is also why it takes longer to make jam. The less homogenous the mixture is, the more time it takes to achieve an even thickness for the perfect spread. Other than that, jellies and jams are made from the same fruits, taste similar, and have almost identical nutritional values, which is why they are interchangeable in recipes.

Making Homemade Jam

The main difference between store-bought and homemade jams is in the quality and the quantity of the ingredients. Making your own jam at home gives you the option to choose how much sugar and pectin you want to add—if any. Some fruits have naturally high sugar content, so you might find that you won't need to add any to your jam, depending on your taste. But keep in mind that not adding extra sugar will make the thickening process even slower. And if you add pectin to thicken your jam, you will definitely need the sugar as well. This is because the bond between the sugar and the pectin will give your jam the perfect consistency.

What is Pectin?

Pectin is the natural fiber in fruits that, when combined with added sugar, helps the liquid inside a jam or jelly to reach a thick consistency, which makes them spread evenly on bread. However, this doesn't mean you have to add pectin to your fruit while cooking when making jam. You can learn how to make jam with or without pectin.

Making Jam with Pectin

The good thing about adding pectin to your jam is that you won't have to worry about the type of fruit you choose. While tart fruits are high in pectin, sweeter ones have not as much, and without added pectin, it can take ages for the jam to thicken. If you are going with low pectin fruit, such as peaches, apricots, or tropical fruit, you will need to use some pectin in the mixture. You can reduce the amount of pectin you add by choosing ripened fruit and adding lemon or apple juice. These acids can act as boosters for pectin release. You will also have to make sure you add the right amount of sugar along with the pectin, because the pectin cannot work without it. The more pectin you add, the more sugar you will need.

Here is your guide to making jam with pectin:

- **Clean and dice the fruit:** wash the fresh fruit under running water before cutting it into small pieces. After that, put the fruit pieces in the saucepan. Besides the size of the chunks, the type of fruit you use can also determine how long it takes for the pectin to act, so keep that in mind when you cut it up. Aim to cut low pectin fruit into smaller chunks to spend less time making your jam.

- **Combine with sugar:** pectin works best with granulated sugar. Make sure to properly combine it with the fruit before heating it. Powdered pectin should also be added at this point. Using a blender or food processor to break up the sugar and the pectin can make them dissolve much faster. Alternatively, you can use preserving sugar, which will help you to store your jam for longer in your pantry. If you use frozen fruit, your first step will be to let it thaw, and add the sugar afterward to avoid burning it.

• **Boil the mixture:** after melting the sugar, increase the heat under your pan to the highest setting to bring the mixture to a full boil. At this stage, you can mash frozen fruit while stirring it with a wooden spoon. If you use liquid pectin, add it to the boiling mixture and let it boil for at least fifteen minutes. Lower the heat to medium and add an acid, such as lemon juice, stirring the jam often until thickened. With pectin, this shouldn't take more than an hour.

• **Skim and jar the jam:** when pectin reacts with sugar, it tends to make a lot of foam on the surface of the jam. Some of this foam will remain, even after the jam is cooked, and you will have to skim the surface to avoid air bubbles in your spread. While you do that, the jam will probably be cool enough for you to scoop it into sterilized glass jars. If you want to store it in the fridge, let the jam cool down to room temperature and properly seal the jar first. Or you can proceed with a canning method to preserve it for even longer.

Making Jam without Pectin

Instead of adding pectin to your jam, consider using fruits that are naturally high in pectins, like apples, blackberries, cranberries, pears, and plums. The lack of pectin will make the process a little slower, as the jam will need more time to thicken, but it will also make for one less ingredient to worry about. Plus, taking the pectin out of the recipe can eliminate the need for sugar as well. Without all that added sugar, this jam recipe could be a great alternative if you cannot consume refined sugar due to any health reasons. If you have chosen fruit with low pectin and are worried about not creating a thick enough jam, you can add an alternative thickening agent, such as chia seeds.

Here is how to make jam without pectin:

- **Wash and cut the fruit:** before you use fresh fruit for your jam, you need to rinse it under cool, running water right before cooking it. After that, cut the fruit into either halves or fourths and put them in the pan. The size of the pieces will determine how fast they cook, so you want to make them as small as possible. You can even blend the fruit with a food processor if needed. If you are using frozen fruit, you can skip this step.

- **Add a sweetener:** the rule of thumb is to add one tablespoon of sweetener for every cup of fruit you use, but you can use less or more depending on your preference. Although table sugar is the most commonly used sweetener, you can opt for a healthier alternative, such as honey or coconut sugar. With fresh fruit, you can add the sugar or sweetener right after you cut it up. When cooking with frozen fruit, you will have to thaw out the fruit first, or the sugar will burn while cooking.

- **Cook the fruit:** place the pot with the fruit and the sugar over a heat source. Bring the mixture to the boiling point and let it boil for at least fifteen minutes, mixing occasionally. After that, reduce the heat to medium-low and continue to stir until the jam thickens. As the fruit cooks, you will notice some bubbles of foam appear on the top. When the bubbles become smaller, your jam is as good as it gets. Depending on the amount of fruit you used, this can take anywhere between thirty minutes and three hours.

- **Cool down and store:** once your jam reaches the consistency you like, remove it from over the heat immediately. You can now sprinkle and stir in some chia seeds to achieve a thicker spread. Let the jam cool, then scoop it into your sterilized glass jars. If you want to freeze or store the jam for a longer period, you can add a little bit of lemon juice at this point. Cover the jars with a tightly sealed lid, let it cool to room temperature, and then place in the fridge.

-

Making Homemade Jelly

Essentially, jelly consists of the same ingredients as jam: fruit, sugar, and pectin. However, when making jelly, you will need to separate the fruit juice from the pulp. This process removes the added texture and some of the natural pectins from the fruit. Because of this, the use of pectin is a must in jelly-making. Even if you use a high pectin fruit such as a citrus fruit or some berries, you will need the additional pectin to form a solid, translucent spread from your fruit juice.

Here is how to make jelly at home:

- **Prepare your fruit:** wash your fruit in cold water under a tap and cut it up if needed. Put the fruit in a large saucepan along with some water. The amount of water you add depends on the softness of your fruit, but it shouldn't be more than 1 cup per pound of fruit.

- **Boil the fruit:** bring the fruit and water to a boil, then reduce the heat to cook on a medium setting. Stir it occasionally until you notice the fruit has softened, and you crush the pieces to extract the juice. The cooking time depends on the fruit you have used, and it may take anywhere from five to thirty minutes.

- **Strain the juice:** when the fruit is cooked, remove it from the heat and let it cool. Place a cheesecloth over a sieve and strain the cooked fruit and water mixture over another saucepan. Let the liquid seep through naturally, even if it takes a long time.

- **Heat the juice:** mix in two tablespoons of lemon juice with the fruit, and if you are using powdered pectin, add that as well. However, if you are using liquid pectin, you should add it when your juice is already hot as it takes less time to melt. Bring the juice to a boiling point while stirring it frequently.

• **Add the sugar:** lower your heat and add a cup of sugar for every cup of juice. If you opt to add an acid, such as lemon or apple juice, you can reduce sugar by a third. On the other hand, tart fruit juice may require a little more sugar, depending on your preference.

• **Cook until jellied:** continue boiling your juice until it begins to thicken. When that happens, you will need to stir it constantly so it doesn't clump up. Due to the added pectin, the jelly will form pretty quickly and visibly. After it has formed, remove it from the heat source.

• **Jar it while hot:** unlike with jam, you shouldn't wait for your jelly to cool down, as this will make it harder to handle. Skim off some excess foam after it's cooked, but you should scoop or pour it into containers while the jelly is still hot and somewhat workable. Remember to leave the jars open to allow the steam to escape before sealing them with sterile lids.

Tips for Making the Perfect Homemade Jam or Jelly

The Best Fruit to Use

Although you can turn almost any fruit into excellent jellies and jams, the fruit you choose will determine the need for the rest of the ingredients. For example, granny smith apples and blackberries are the best fruits to make jam for beginners. They have higher acid and pectin content than any other fruit, and you can get a lot of jam from them. Raspberries and strawberries are also great for making jam if you remember to combine them with pectin and sugar. For a thick jelly, use plums, peaches, or even cantaloupes. Or you can think outside the box and use dandelion or mint instead. While they technically aren't fruit, they are inexpensive and can add a delicious flavor to your jelly.

Mix and Match

Sometimes the best results will come when you begin to mix flavors. Not only that, but by combining high pectin fruit with low pectin ones, you can further reduce the amount of pectin and sugar that you will need to add. Mixing pears and apples can increase the overall pectin content while counteracting the tartness of your jam or jelly. Adding pomegranate to your blueberry jam can act as a strong emulsifier, plus the flavors come together wonderfully. Berries, in general, blend well together when used to make jelly. Besides fruit, you can add some other food elements to your jams and jellies. Combining oranges with ginger or adding a small number of hot peppers into a jelly can bring joy to your taste buds and ease your digestion.

Avoid Artificial Sweeteners

When used in a large quantity, artificial sweeteners can actually turn bitter, something you definitely don't want for your sweet fruit spreads. And even if a sweetener gives you the taste you want, the pectin cannot bond to it as it will to the sugar, and your jam or jelly can end up runny. This can happen with some natural sweeteners as well, so be careful when choosing.

Chapter 7: Pressure Canning

Pressure canning is used to safely and effectively can low acid foods like fruits, vegetables, soup, and meat. Due to the increasing popularity of multi-utility pressure cookers, many people believe that pressure cooking can be used for canning. However, that is not the case. The process of pressure cooking might be similar, but it isn't the same as pressure canning.

You'll need a pressure canner like the Presto 16-Quart Aluminum Pressure Canner to successfully increase your food's shelf-life. Unlike high-acid foods like jam and pickles, low-acid foods are prone to more potent bacteria. Simple water-bath canning won't be sufficient for canning your batch of beans, soups, or veggies safely.

Pressure canning can be frightening, especially if you're a beginner. This chapter underlines the science behind it, the differences between water bath canning and pressure canning, and the importance of proper equipment to make the process less intimidating. It also talks about the types of products that can be preserved, step-by-step instructions to operate a pressure canner safely, and things to avoid while using the canner.

What Is a Pressure Canner?

Like a pressure cooker, a pressure canner is a large pot that can be locked shut with a lid. High pressure is built up inside the container when intense heat is applied to it through a burner. A pressure canner generates higher boiling temperatures by increasing the pressure. By processing low-acid foods in a pressure canner, they can be sterilized and stored in your cabinets for a longer duration of time.

Pressure canners have a regulator. The regulator can either be a one-piece pressure regulator, dial-gauge regulator, or weighted-gauge regulator. The weighted gauge or dial on a pressure canner lets you control the amount of steam generated inside the canner simply by turning the burner heat up and down. All pressure canners have a knob or dial-like regulator that allows you to control the pressure inside the canner.

How Do Pressure Canners Work?

A pressure canner uses pressure to create temperatures that are way higher than boiling temperature. This helps to heat and process low-acid foods effectively to last longer when sealed and stored appropriately. If you're using a weighted-gauge pressure canner, you can set a weight on the steam valve corresponding to a pressure of five, ten, or fifteen PSI. On passing the set weight, the pressure canner releases a small amount of steam to stabilize the pressure. However, a weighted-gauge pressure canner only gives three weight settings. On the other hand, in a dial-gauge pressure canner, you get more options to control the pressure and maintain an intermediate temperature level.

The Difference between Water Bath Canning and Pressure Canning

There are two types of canning methods. The first is called Water Bath Canning, which does not require any special equipment, except for the canning jars. However, the pressure canning method requires a specialized device known as a pressure canner. You must understand that both these methods are utilized for different purposes. That said, each of these methods has different procedures to be followed for different types of foods. If you use the correct method and procedure, you can easily and satisfactorily preserve your food.

The process of water bath canning involves a large pot, usually filled with boiling water. Tightly sealed jars of food are completely immersed in the boiling water bath for the required amount of time, as specified in the canning recipe. The jars are then allowed to cool down until a vacuum seal is created. A boiling water bath can only heat the food to the boiling temperature and is well suited for acidic foods like pickled vegetables, fruits, and preservatives.

The pressure canning method involves a pressure canner with features like a pressure valve, vent, and screw gauges. This method is perfect for canning less acidic foods like certain vegetables and meat with a near-neutral or alkaline pH level. Apart from the equipment used, the main difference is that food can be heated up to higher than boiling temperatures if you use the pressure canning method.

The Importance of Proper Equipment

It is critical to know that. To process your food safely, you must choose the appropriate canning method for the type of food you want to preserve. Low-acid or alkaline foods like animal products, soup stocks, and unpickled vegetables cannot be processed in a boiling water bath. You'll need a pressure canner because alkaline foods are more receptive to bacteria that can only be sterilized at temperatures higher than the boiling temperature.

Remember this: all alkaline foods must be processed in a pressure canner. In contrast, all acidic foods like pumpkin, tomatoes, and pickled vegetables can be safely processed in a boiling water bath. It is also important to note that you should use glass jars instead of metal cans for domestic-use food preservation.

Foods that Qualify for Pressure Canning

As mentioned before, foods acidified using vinegar or citric acids, like pickled vegetables, do not require pressure canning. While all types of foods that have an alkaline or low-acidic pH can be safely processed in a pressure canner. Here's a list of foods that qualify for heavy-duty pressure canning.

- **Beans:** all types of beans like baked beans, black beans, green beans, navy beans, and pinto beans can be preserved by pressure canning.

- **Meat/seafood:** all kinds of meat like seafood, poultry, beef, chicken, and pork can be processed in a pressure canner.

- **Broths:** both meat and vegetable broths and stocks have low acid levels that qualify them for processing in pressure canners.

- **Soups:** stews or soups made with low-acid ingredients like onions, mushrooms, and peppers can be preserved by pressure canning.

- **Vegetables:** pumpkin, potatoes, carrots, sweet corn, and other veggies can be chopped and preserved for long periods by pressure canning.

Supplies Needed for Pressure Canning

If you're trying to increase the longevity of your low-acid foods, pressure canning is the best method to use. But, to do that, you'll need some supplies. First of all, you'll need to purchase a pressure canner, which can cost you anywhere between $100 and $500, depending on the quantity of food you want to be able to process per batch. Plus, you'll need some canning jars, lids, bands, and other canning accessories, like jar lifters, canning filters, ladle, potholders, and kitchen towels. One key requirement for pressure canning is a traditional or gas stove with coil heating units. You'll also require a chopping board, countertop space for preparing food, and a place to keep the empty jars. Once you're done processing and canning the food, you'll require shelf space to store your delicious creations.

Step-By-Step Instructions for Pressure Canning

Here are some simple steps you need to follow to safely and effectively process foods by pressure canning.

1. Pre-Canning Check

Before you start the pressure canning process, ensure you've gathered all the materials that you're going to need. This includes freshly washed glass jars, a canner lid, canner trivet, and other canning accessories. Ensure there aren't any cracks or leaks in the canning jars and that they properly fit the lids.

2. Preparing Your Foods

Depending on the food you're trying to process, some recipes require you to pre-cook the food to preserve its taste and texture. At the same time, some instruct you to directly add your food into the canning jars and place it in the pressure canner. The best way to ensure the proper canning of your food is to follow the instructions provided in tried and tested canning recipes.

3. Pre-Heating the Pressure Canner

The water level in a pressure canner needs to be enough to fill the canner's chamber with steam and prevent the pressure canner from running dry during the process. Start by adding water to the canner as per the device's instructions. Add the canning trivet to prevent your jars from coming into direct contact with the heated bottom of the canner. You can then begin pre-heating the pressure canner until the temperature rises to 140 F - 180 F. It is important to note that the canner's temperature must be similar to that of the jars. Otherwise, you risk breaking your jars due to thermal shock.

4. Loading the Canner

While your pressure canner pre-heats, you can start preparing your food and then fill your canning jars using a ladle and funnel. Ensure you have enough headspace in your jars as per the canning recipe. Tightly turn the lids of your jars with your fingertips. Make sure the lids are tight enough that no food can leak into your canner. At the same time, however, the lids should be loose enough for air to escape when the jars cool down and create a vacuum. Now you can load your jars on top of the canning trivet in your pressure canner with a jar lifter.

If your jar contents leak into the pressure canner, this can be because of too little headspace, or the lids are screwed on too tightly.

5. Sealing the Pressure Canner

Once you've loaded your food jars in the pressure canner, leave the vent open and turn on the heat. Let the entire chamber of the canner be filled with steam. Wait until you see a constant flow of steam coming out of the vent for at least ten minutes before closing the vent. If you fail to do this properly, your canner can be left with cool air pockets. This will give improper canning results and partially unprocessed food.

6. Maintaining the Pressure

Bring the canner's pressure up to the required level and start the timer once that level has been reached. Maintain the recommended pressure from your canning recipe for the required amount of time. When the time is up, turn off the heat and leave your pressure canner as it is for some time. This is to let the pressure canner and the contents within cool down. Avoid rapid cooling of the canner because it can break the jars inside.

7. Release the Steam and Remove the Jars

Check the pressure on the gauge of your pressure canner. If it has come down to 0 PSI, there's no more increased pressure in the canner, and you can safely release the steam through the vent before opening the lid. Safely remove the jars and place them on kitchen towels to cool on your countertop.

Be careful, as the contents of the jars will still be pretty hot for a long time, so you best let the jars cool off overnight. Check and double-check the seals on your jars. Remove the canning rings if they're still on. You can then store the sealed jars in your pantry to be used within 12-18 months.

Things to Avoid

Pressure canning can go very wrong very fast if you overlook any of the safety instructions or forget to follow a step. There's a huge margin for error in the process of pressure canning. To summarize the most common safety precautions, here's a list of things to avoid while operating a pressure canner.

1. Changing the Temperature Too Quickly

The temperatures of the canner and the food being placed in it must be regulated to match one another. If the jars you place inside the chamber are cooler than the canner, they can crack from the sudden temperature change.

2. Rushing the Process

Pressure canning requires that you invest a lot of time if you want to execute the process properly. There are a lot of instructions to be followed and precautions to be taken. If you're impatient and try to rush the process, things will most definitely go unpleasantly. Give the process the necessary time to avoid any complications with canning and to ensure your personal physical safety.

3. Leaving No Headspace

It is necessary to leave the recommended amount of headspace for the contents in your canning jars. If there's not enough headspace, the boiling contents of the jars will hit the lids and cause them to loosen. This will allow liquid to flow out of the jars. You may even experience excessive liquid loss, where more than half of the jar's contents leak out.

4. Releasing Pressure Manually

Unlike pressure cookers, you cannot let off steam within a heated pressure canner manually. Do not try to release the canner's pressure in any way if you've just turned off the burner. Doing so may result in the jars inside the pressure canner breaking. The super-hot steam can also cause severe burns if you try to release the pressure manually.

5. Using an Instant Pot for Canning

Multi-utility appliances like the Instant Pot must not be used for pressure canning. There are two reasons for this. The first reason is that these countertop appliances cannot maintain the even pressure required for safe canning. The second reason is they don't support canning trivets which are crucial to keeping the jars safe from the heated bottom of the pot. As a result of thermal shock, glass jars are bound to break in an Instant Pot.

Tips for Perfect Pressure Canning

Are you canning your foods just to preserve them longer or to preserve them in a way that maintains their taste and texture? Pressure canning, if performed according to instructions, can be an extremely satisfying experience. Here are a few tips to help you preserve your delicious foods perfectly.

1. Learn to Use the Pressure Canner

Read the instruction manual of your pressure canner carefully before using it. Understand its functionalities and requirements. If you have a dial gauge pressure canner, have the dial inspected for accuracy every 8-12 months. If your pressure canner hasn't been used in a long time, get it inspected for leaks.

2. Inspect the Jars

Clean your canning jars with hot water and soap. Rinse them well. Check for cracks or nicks on your jars. Make sure the lids fit perfectly on the jar heads. With a spatula, release the air bubbles between the jar and the food, if there are any. As per directions, leave a headspace of ¼", ½", or 1" in your canning jars. Always use a jar lifter when placing the jars in boiling water or removing them from the canner. Instead of using old-fashioned jars or used peanut butter jars, try mason-style jars to store your canned foods.

3. Label Your Canned Food

Once you're done pressure-canning your delicious recipes, label the jars with the content name and date. Don't forget to mention the canning year on the label. It may seem silly to some, but it's easy to lose track of the food stored in your pantry. Labeling the jars will help you to identify the ingredients of the jars and whether or not they have expired. It's easy to label your canned food as soon as it cools down and very helpful when you stumble upon a canning jar far in the future!

Chapter 8: How to Can Fish

Give a man a fish, and you feed him for a day; teach a man to can fish, and he can make it last for an entire year!

Canning is a safe and easy preservation method for fish that has been around for hundreds of years. In fact, the first reported instance of canning was done by Nicolas Appert—the French inventor who is also known as the "father of canning"—who invented this method solely to preserve fish. The science behind canning fish is quite simple. By cleaning the fish and packing them in disinfected, air-tight jars, you slow down their spoilage. This environment prevents the growth of microorganisms such as bacteria and mold, which keeps the clean pieces of fish safely edible for up to a year.

Apart from being convenient and cost-effective, canned fish has many health benefits that even rival fresh fish at times. They have high protein content, are rich with omega-3 fatty acids, and tiny, tender fish bones that you can safely eat, making canned fish one of the best sources of pure calcium out there. Therefore, it is quite common to can the highest quality yields to be consumed when the weather is too cold for fresh meats and food, especially in countries with really cold winters. It is more or less similar to cured meats that make the end product different, delicious, and more valuable than fresh meats.

Types of Fish Ideal for Canning

Fish are relatively easy to preserve, some being easier than others, and they keep their flavors and nutrients well after the canning process. However, no matter which type of fish you plan to can and preserve, it is important that you select top-quality, fresh fish. Like any other fresh meat, fish can be susceptible to tissue decomposition if not handled and preserved well. Always use freshly caught fish for canning. If you use wild-caught fish, try to keep them alive for as long as possible if you plan to preserve them later. They start the deterioration process soon after they leave the water. Once you have them, clean and preserve them as soon as possible to create the best-canned fish. Here are a few species of fish that are ideal for canning.

Mackerel

Mackerel is an affordable fish with a taste similar to tuna and salmon. They are easy to catch and can be found near piers and rocks in large flocks, ideal for preserving. They are rich in healthy fats and have a decent number of bones that become tender during canning. While fresh mackerel can be somewhat chewy, canned mackerel is tender and has a distinct yet mild flavor.

Salmon

Salmon is a popular fish for canning because of its delicious taste, high omega-3 fatty acid content, and potent antioxidants that give its uniquely appetizing color. Wild-caught salmon are often the best choice for canning since they have fewer contaminants and are more nutritious. Once properly canned, salmon has a light color and texture, with a pleasantly mild flavor.

Herring

Herring are small, bony, and naturally oily, making them great for canning. The excess moisture in the fish escapes and combines with the salt and makes for a deliciously tender and flavorful meal when you cook and eat them later. Due to their size, flavor, and moist texture, canned herring takes little to no preparation or cooking.

Trout

Trout is a member of the salmon family and is a popular fish among hobby fishermen since they are widely available in creeks and rivers. Fresh trout has tender and flaky flesh with a mild nutty flavor. They have ample flesh and can be smoked or brined before canning to add more flavor if needed.

Steelhead

Steelhead is also a type of salmon with a nice orange flesh and milder taste. A great source of low-fat protein, steelheads are anadromous fish that spend parts of their lives in freshwater and others in saltwater. The flesh is tender and moist, making the canned fish flavorful with great texture.

Blue

Bluefish are large marine fish with mild, flaky flesh packed with intense flavor. They make rich and succulent canned fish, often tasting even better than fresh counterparts since the preservation makes the strong flavor milder and more pleasant. Due to their ample flesh, bluefish give a great yield as canned fish, which can be used to make plenty of food, including fish cakes, soups, pasta sauces, or croquettes.

How to Make Great Canned Fish

Now that you have your fish—fresh fish is always the best option, but thawed frozen fish can also be canned—it's time to gather the equipment and ingredients and start the canning process.

Equipment

- Filleting knife.

- Hardwood cutting board.

- Wide mouth canning jars and self-sealing lids with rings.

- Pressure canner.

Note: As mentioned in the previous chapter, a pressure canner is a big pot with a secure lid, with a dial or a gauge that helps you regulate the steam pressure. Be careful when you use the pressure canner since the pressurized steam can be hotter than boiling water. Every pressure canner on the market comes with its own set of directions. Read your manual several times and follow the exact directions when you use it. Before every fish canning session, check if your pressure canner is in good condition. The vent should always be clean and open, and the pressure gauge should be accurate. Doing a trial run with a few inches of water to check the performance of the pressure canner is recommended before every canning session.

As for the jars, you can use straight-sided mason-type canning jars with a tight lid. If you are using jars that have been used previously, thoroughly clean them using hot, soapy water to remove any residue. For your preserves to retain their quality, the lid should be perfectly secure on the rim of the jar. Check for any nicks or cracks on the rims and discard any damaged jars. Get self-sealing lids to secure the jars and replace them if you use them for a second time. You can reuse jars and rings if they are not damaged or bent.

Every piece of equipment used in the canning process, including the knives and the hardwood cutting board, should be cleaned well with warm, soapy water, rinsed, and properly dried.

Ingredients

- Fish
- Salt
- Water
- Vinegar (optional)

Preparation

If your fish is frozen, put it in the refrigerator until it is completely thawed before getting started. Rinse fresh or thawed fish well in clean, cold water. You can add some vinegar to the water to help remove any slime. Gut the fish and remove the head, fins, scales, and tail without damaging the skin or flesh. You can keep the bones in since the pressurization tenderizes the bones, making them a good source of calcium. Use a generous amount of water to wash and clean the fish that you will be canning. If you are not going to start the canning process immediately after preparing the fish, store it in a refrigerator until you are ready.

Packing

Clean and disinfect your canning jars. Cut your fish into chunks that can fit into the jars. If you are using a type of fish where the skin stays on, arrange the pieces in the jars with the skins facing outwards to make the finished product look nicer. If the fish skins are touching the insides of the jar, it can be difficult to clean after use. You can simply face the skins inwards if you want to avoid the few extra minutes of cleaning after finishing the jar. If you are using standard 1-quart canning jars, pack the chunks of fish tightly, leaving about one inch of unfilled space at the top. Use a plastic spoon to align the product inside for a tighter pack. If you prefer some added flavor, add about one to two teaspoons of salt to each jar. You can also add small amounts of your favorite herbs and spices, such as paprika and garlic powder, to the jars as well. When you can halibut, add a couple of spoons of olive or vegetable oil for better moisture.

Once you pack the fish in, use a clean, damp cloth or a paper towel to remove any oils or residue from the edge of the jar. Secure the lid and the rings properly. Read and follow any specific guidelines from the manufacturers of your canning jars and lids. They should neither be too tight nor too loose. If you over-tighten the lids, it can lead to the jars breaking and discoloring the fish, since air will not escape through the tight lids during processing.

Processing

Add water to the bottom of the pressure canner until it is three inches deep. Put the rack on the bottom and place the closed jars on the rack. Check any specific instructions in your pressure canner manual on arranging the jars for more efficient processing. Fasten the canner cover. The directions for processing fish in the pressure canner vary according to the jar size, fish type, and canner brand. Always check the instructions from the manufacturer. For standard quart jars, heat the canner on high for about 20 minutes until you can see steam coming through the open vent in a steady stream. If the steam is not steady enough, let it heat up a little more. Let the steam flow out for about ten more minutes. This makes sure that the heat spreads evenly inside the canner. The time it takes for a steady stream of steam to come out may be thirty minutes or more, depending on the size of your canning jars and the temperature of the fish.

When the heat becomes even, close the lid vent using an oven mitt or hot pad and set the weighted gauge. Depending on the manufacturer, there should be three sections in your weighted gauge. Turn up the heat until the pressure reads 10 pounds for a weighted gauge pressure canner. Keep adjusting the heat to maintain a steady pressure from the beginning to the end. If you are using quart jars, it may take about 160 minutes of processing with 10 pounds of pressure for most fish. If you are located at over 1000 feet altitude, use up to 15 pounds of pressure.

Cooling

Once the recommended processing time is over, remove the canner from the heat and let it cool. Allow the pressure to drop naturally until the gauge shows zero pounds of pressure. Most pressure canners nowadays come with a lid lock that automatically unlocks when the pressure drops to zero. Wait a few more minutes and slowly open the vent using a heating pad or oven mitt. Open your lid so that the escaping steam faces away from you. Take the jars out one by one using tongs or a jar lifter. Do not try to tighten the lids if they appear to be loose. It takes some time for the lids to seal while cooling. Check the lids after about twelve hours to see if they have been sealed properly. You can remove the rings and wash them if you want to reuse them later. If a lid has not sealed after twelve hours, remove the fish and use a different jar to process them within twenty-four hours. You can place the unsealed jars in the freezer until you can process them again.

Storing

Wash the surface of your sealed jars and wipe them dry. Label them with the processing date and the type of fish, if you are canning more than one type. Store the jars in a cool, dry place.

Chapter 9: Canning Poultry and Meats

Poultry and meat canning is easy if you know how to do it. You won't find it much different from canning veggies and other vegetarian foods. Not only does canned meat and poultry stock your pantry, but it allows you to prepare quick meals when needed. Imagine the convenience of having a canned jar ready to be opened to add in to one of your delectable meat or poultry recipes, without the need to prepare the meat or chicken.

Besides convenience, canning meat or poultry products is a must-have skill that will save you space in your freezer and keep you prepared with ready-to-go food that you are sure to cherish.

However, canning poultry and meats is only healthy and possible with pressure canning because these foods have very low pH levels (less than 4.6). Using water baths or steam canning will not be enough to heat the food in the can, making it unsafe for storage.

Many of you may feel intimidated by the process of pressure canning, but following the below instructions will help you to secure your food safely for years.

Canning Chicken

Every poultry product that you want to can will require the same steps, so let's understand the process by knowing how to can chicken. You need to first do the preparation work for the chicken-canning recipe.

Preparation Work before Canning Chicken

Store-bought chicken is seasoned and refrigerated, so you can use it as soon as you buy it. However, if you plan to use freshly butchered chicken, season it first. Then, let it chill for six to twelve hours before you start canning it. Make sure you thaw the chicken pieces before you use them if they were frozen.

Next, choose the canning options you want to use for the chicken.

Hot or Raw Pack?

In the hot pack option, you first cook the chicken lightly and then put it into canning jars with some of the hot liquid and preserve it. The raw pack method requires you to put raw chicken chunks into the jars and preserve them. While both methods are safe to preserve chicken, the hot pack one makes the chicken last longer.

Note that you don't have to cook the chicken too much because pressure canning will do that for you. That's why it's more convenient to go for raw pack canning.

Bone-in or Boneless?

It's your choice whether to keep the bones intact in the poultry meat or not. Either way, it will not make any difference to the quality of the canned chicken. However, a boneless chicken will give you more space for food. In addition, chicken pieces with bones will require you to use more effort to fit them inside the jars properly. Using boneless meat will give you the advantage of storing somewhat uniform chicken cubes in the jars.

Equipment and Ingredients for Canning Chicken

It's always a good idea to start canning food after cleaning the kitchen counter, preparing the area, and setting the equipment up. Canning requires adequate space to keep all the tools, jars, and other equipment neat and tidy to ensure maximum hygiene.

Here's what you'll need to have ready before you start the canning process.

- Chicken (boneless or bone-in, certain parts or the full chicken chopped into chunks).

- Salt (completely optional, most people add it to enhance the flavor of the poultry meat).

- Canning utensils.

- Canning rings, lids, and jars (pint or quarts – depending upon your canning requirements).

- Pressure canner.

- Clean rag.

If you want to can chicken for one meal, using pint-sized jars may be the best choice. However, if you don't mind using leftovers, quart could work too. Now, let's look at the recipe.

Chicken Canning Recipe

Prep Time: 30 min

Cook Time: 90 min

Total Time: 2 hours

Directions:

1. Start by preparing the pressure canner. Pour water in the canner up to several inches, and start heating the water at low heat.

2. Chop the chicken (bone-in or boneless) into small pieces to make them convenient enough to fit in the jars. If you like, you may remove the chicken skin. As mentioned earlier, it will be easier to fit boneless pieces in the jars, as pieces with bones will require some adjustment.

3. For the raw pack method, fill the jar with the chopped chicken pieces in raw form. You may add 1 tsp. of salt for quart jars or ½ tsp. of salt for pint jars to enhance the flavor.

4. If using the hot pack method, steam, bake, or boil the pieces very lightly (2/3rd of the way cooked) before adding them to the jars.

5. In both cases, pour hot water or broth over the packed chicken pieces. Make sure to leave 1 inch of headspace between the rim and the poultry.

6. Use the air-bubble removal utensil, a chopstick, or plastic knife to remove the air bubbles trapped among the chicken pieces. Avoid a metal knife as it can scratch the glass jars and damage them.

7. Use a clean cloth to properly wipe the residue from the jar rims.

8. Place the lid and close the ring over the jar. Ensure that the lid is just closed finger-tight.

9. For jars with boneless chicken, set the canning process to 75 minutes for pint jars and 90 minutes for quart jars. For canning bone-in chicken, set the process to 65 minutes for pint jars and 75 minutes for quart jars.

10. If you are using a dial-gauge pressure canner, can at 11 lbs. or 12 lbs. of pressure. If using a weighted gauge canner, can at 10 lbs. of pressure or at 15 lbs. of pressure.

Now, let's look at a recipe for canning meats other than poultry.

Meat Canning Recipe

Prep Time: 1 hour

Cook Time: 2 hours

Total Time: 3 hours

Directions:

1. Make sure you are using chilled meat. Properly thaw the meat if frozen. You can either use a microwave or cold water to thaw it.

2. If the meat is strong-flavored, soak it in brine (1 tbsp. of salt per quart jar of water) for 1 hour. Rinse the meat and remove excess fat. Chop the meat into chunks, cubes, or wide strips.

3. If you use the hot pack method, make sure to precook the meat by browning, stewing, or roasting it in a little bit of fat. Don't cook it completely.

4. Place the rack in the pressure canner. Pour water in the pressure canner to a few inches, and get it boiling.

5. Pack the prepared meat or raw meat in the jars cleanly, leaving 1 inch of headspace at the top of the jar. Make sure the meat is not packed too tight in the jar because you want the liquid to flow freely around the meat.

6. Add salt to the jars (1/2 tbsp. to pint jars and 1 tbsp. to quart jars) if desired.

7. Pour a few inches of boiling water into each jar.

8. Use a clean rag to wipe the rims and make sure they are dry. Run your finger around the rim to remove knicks or salt particles, because such elements can disrupt the sealing process.

9. Affix the lid and rings on the jars and place them into the pressure canner.

10. Place the jars on the rack so that they are a few inches above the base of the pressure canner to prevent damaging the jars.

11. Set the pressure canner to start building pressure, keeping the gauge off. Allow the canner to build up steam and vent out for 10 minutes before placing the weighted gauge.

12. After the steam has vented out, keep the pressure building until the dial gauge shows 10 lbs. of pressure when operating at sea level. If you are located at 1,000 feet or more above sea level, let the pressure in the gauge increase up to 15 lbs.

13. For pint jars, keep the pressure for 75 minutes. For quart jars, retain the pressure for 90 minutes. After the appropriate time passes, allow the canner to release the pressure naturally.

14. Once the canner releases the pressure, open the lid, facing away from the steam that will be released. Use canning tongs to vertically pick up each jar.

15. Place them on the counter and let them cool and seal for 24 hours. Once sealed, you can remove the rings and store the sealed jars as needed.

Note: Instead of chunks, strips, or pieces, you can also use ground meat for the process. You should only use a pressure canner for meat and poultry, or it will not work properly. As mentioned before, low-acidic foods (with pH lower than 4.6) require a high level of heating to ensure that the bacteria is killed off completely. Otherwise, it can lead to botulism.

How to Practice Safe and Efficient Meat and Poultry Canning

There is no doubt that pressure canning is a convenient, efficient, and quick way to store meat and poultry for later consumption. It saves you the trouble of going to the store to buy ingredients, wasting a lot of your time. Additionally, it's a convenient way to keep food with you in case of an emergency.

However, if you don't follow the steps for pressure canning carefully, you may end up damaging your health. That's why it's important to follow the below-mentioned tips for canning poultry and meat safely and efficiently.

Make Sure You Are Using the Right Equipment

It's important that you only use a pressure canner for processing meat and poultry products. Don't use a steam canner or water bath canner for these foods. The pressure canner should have a high-quality weight or dial gauge. Next to the canner, ensure that you have new jar lids. Follow the safety rules and protocols listed in the pressure canning manual to properly use the equipment.

Prepare for a Large Batch

Canning meat can take a lot of time, so make sure you have enough room for a large batch. That way, you can save time and effort and avoid the need to can frequently. If you have to buy a pressure canner first, make sure to invest in a large model that will allow you to process several jars at a time. Using a large canner will help you to increase your efficiency.

Don't Be Afraid of Using a Pressure Canner

Many people are afraid of pressure canning because it feels like the equipment may explode any minute. However, that's not the case at all. If you follow the simple instructions for using a pressure canner, you won't have any problems. After one or two canning batches, you will find it to be the most reliable equipment for preserving meat in the long term.

Enhance the Meat Flavors

If you want the canned chicken or meat to have a good flavor, use dried herbs or salt when processing the meat in jars. Herbs and spices tend to improve the flavor of food if preserved for a long time. One of the best examples is pickles, which a lot of people enjoy eating. You may also use green herbs, but that usually changes the meat color to a shade of green, which may not look very appetizing. If that's the way you see things, too, go for dried herbs instead.

Preserve Strategically

Make sure you know the right quantity of meat that you want for the canning process. If you or your family don't consume a lot of meat, it will be useless to can and store a lot of canned meat. In such a case, you may just use the freezer to preserve the meat or chicken and utilize the canning space for other foods.

Can What You'll Use

Preserve meat that you will be using in your dishes later. Make sure to know what you and your family consume the most. It would be useless to preserve beef chunks if you only eat white meat, for example. Likewise, preserve meat in the right form, such as chunks, strips, or ground, as per your liking to add to specific dishes later.

Prepare Each Jar in the Right Quantity

Most families have an idea about the average amount of food each member consumes. Keep that in mind when selecting the canning jar sizes to have enough meat for the meals you will make later. For instance, use a 1½-pint or wide-mouthed jar to store loaf-style meat or bologna so that you find it easy to take out the meatloaf as per your needs.

Keep the Rings Affixed

Most canning experts advise removing the ring from canned jars containing pickles, jellies, or jams when storing them. This prevents the rings from rusting or becoming infested with insect nests. However, you may want to keep the rings intact on jars with canned meat to add an extra layer of protection against rodents. This depends on where you store your jars. Also, ensure that the jar lids and rings are completely dry before storing them.

Cooking Meat before Canning is Optional

Most people preserve meat after cooking it a little to be ready for specific recipes when opened. However, it's completely optional to do so. You can use the raw-pack method for canning poultry and meat without any issues. If you want, you can add spices or salt to enhance the flavor of the raw meat.

Monitor the Meat Jars before Consuming

Even if you have successfully followed the steps of pressure canning meat or poultry, you should still habitually check the stored jars for any signs of inadequacy. Discard the meat immediately if you notice anything unusual. When you plan on using the meat for a recipe, make sure to reheat the meat and stock by boiling the ingredients at 140° F for at least 10 minutes. Do this even if the preserved meat in the jar appears to be completely fine. This removes any possibility of infection in the preserved food, making it safe for consumption.

Remove the Jars in a Vertical Position

It's important to keep the jars upright whenever you handle them so that the pressure built up in the jars with the meat can be sealed efficiently. Tipping the jars or placing them unevenly can compromise the sealing process.

Pressure canning sure does need practice, but the benefits of doing so can be fruitful. It's important to follow the necessary safety tips to ensure maximum efficiency when preserving or using the food. Unlike other foods, meat and poultry items require extra care because they are not acidic, making them more vulnerable to bacterial growth if prepared carelessly. Note that using a pressure canner is not as challenging as preserving the meat safely in jars. Pressure canning is a safe and fun process that will let you keep food for months and years without worrying about shortages at the time of need.

Chapter 10: How to Ferment

For thousands of years, even before developing alcoholic drinks, humans have been, unknowingly, fermenting food. The fermentation process of dairy was likely a natural occurrence because of the innately present microflora and the hot climate. Researchers even suggest that hanging goat milk bags over the backs of camels was the world's first yogurt production process. It wasn't until 1856 that the science behind fermentation was understood. That year, Louis Pasteur, a French chemist, linked yeast to the fermentation process.

Later in 1910, Elie Metchnikoff, a Russian bacteriologist, brought new information regarding fermentation to light. He suggested that since Bulgarians consumed more fermented dairy than other nations, they had a longer average lifespan of 87. His observations suggested that fermented food is considered beneficial to human health. Further investigations revealed that Lactobacillus acidophilus, the bacteria found in fermented dairy, survives inside the human gut and remains very active. Throughout the 1900s, fermentation was used popularly as a food preservation method. By storing food in an oxygen-free environment, they were able to keep food from spoiling. Undesirable bacteria can't survive anaerobic environments, while desired bacteria thrive.

For the past forty years, give or take, considerable research regarding the health benefits associated with the consumption of "good" bacteria has been conducted. Several links were made between benefits, including detoxification and improved digestion, and the consumption of friendly bacteria. You may have heard about the endless probiotic products available, from supplements to beverages, which have become popular in today's health and fitness world. Probiotics are a commercial trend right now, and it's nothing to be upset about. Fermentation comes with great benefits, donates a very strong, unique flavor to food, and is a great way of preserving food. That is why it is no surprise that you are interested in learning how to ferment your food. One thing to keep in mind before jumping right in is that fermentation is substantially modulated decay.

What Is Fermentation?

In simple terms, fermentation is a metabolic process where the activity of microorganisms results in a change in food and drinks. This change is usually desirable as it is used to add flavor, increase health benefits, preserve food, and more. Although the word "ferment" is derived from the Latin word "fervere," which means "to boil," the fermentation process can occur without the presence of any heat.

How Does It Work?

Good bacteria survive by feeding on carbohydrates for fuel and energy. Adenosine triphosphate (ATP) and similar organic chemicals transport this energy to each part of the cell whenever needed. ATP is generated when microbes respire—they can produce ATP most efficiently through aerobic respiration. When glucose is converted to pyruvic acid—a process called glycolysis—aerobic respiration starts but on the condition that sufficient oxygen is present. On the other hand, fermentation is a process that's similar to anaerobic respiration, which occurs without the presence of oxygen. The production of ATP is also possible in such an environment. This is because the fermentation process results in lactic acid production and other various organic molecules that result in ATP. In this case, good bacteria also feed on carbohydrates, starches, and sugars, releasing carbon dioxide, alcohol, and organic acids, which preserve the food and give it flavor. Individual microbes and cells generally can alternate between the two energy production modes, depending on the surrounding environment.

What Goes on in the Process?

As mentioned above, fermentation is an anaerobic process when oxygen is absent and good microorganisms, like bacteria, yeast, and mold, are present to acquire energy from fermentation. In fact, some yeast cells, like Saccharomyces cerevisiae, favor fermentation over aerobic respiration when enough sugar is present, even in the abundance of oxygen. As the fermentation process takes place, good microbes break starches and sugars down into acids and alcohol. This preserves food, allowing us to store it for prolonged periods without it spoiling.

The enzymes that fermentation provides are also vital for digestion. Humans are born with a specific number of enzymes, and as we age, they decrease. Fermented food provides us with the enzymes necessary to break foods down. Fermentation helps with pre-digestion as well. As microbes digest starches and sugars, they break down the food before we even consume it.

Types of Fermentation

There are generally three types of fermentation that you can use: lactic acid fermentation, ethanol or alcohol fermentation, and acetic acid fermentation. In the lactic acid fermentation process, the bacteria and yeast strains convert sugars and starches into lactic acid, requiring no heat. Since this is an anaerobic chemical reaction, pyruvic acid uses the nicotinamide adenine dinucleotide + hydrogen, or NADH. A similar process is carried out by human muscle cells during strenuous activity. Lactic acid bacteria are necessary for producing and preserving food, especially wholesome and inexpensive food.

Impoverished nations rely heavily on this fermentation method. Kimchi, sourdough, pickles, sauerkraut, and yogurt are all made using this fermentation method.

During the ethanol or alcohol fermentation process, pyruvate molecules and the product of glycolysis are broken down by yeasts. Carbon dioxide molecules and alcohol are the products of broken-down starches and sugars. This fermentation method is used to make wine and beer. The acetic acid fermentation process ferments the sugars and starches of fruits and grains into vinegar and other condiments with a sour taste. Wine vinegar, apple cider vinegar, and kombucha are examples of the resultants of this process.

Stages of Fermentation

The stages of the fermentation process can vary according to what you are fermenting. However, there are generally two fermentation stages: primary fermentation and secondary fermentation. Primary fermentation is a relatively brief phase, during which the microbes quickly begin to work on raw ingredients like dairy, vegetables, or fruit. The surrounding liquid, such as the fermented vegetables' brine, contains microbes that prevent food colonization from the putrefying bacteria. Carbohydrates are converted into acids and alcohols by yeasts and similar microbes.

During the longer fermentation stage, or secondary fermentation, carbohydrates, the microbes' and yeast's food source, becomes scarcer, causing them to die off. The alcohol levels rise as well, and this phase usually lasts anywhere between several days and weeks. Alcoholic beverages are made by winemakers using secondary fermentation. The chemical reactions between the microbes and their environment are highly affected by the varying pH of the ferment—the pH can change drastically from what it initially was. Once the alcohol levels reach 12% to 15%, the yeast is killed, and no further fermentation occurs. Distillation is then needed to get rid of the water and condense the alcohol to increase the concentration.

Benefits of Fermentation

There are many benefits associated with fermented products. One of the key benefits is that digestive, health-friendly bacteria in the gut can be restored by the probiotics generated in the fermentation process. This may help reduce numerous digestive problems. For instance, a study from the *World Journal of Gastroenterology* suggests that probiotics can help relieve Irritable Bowel Syndrome (IBS) and reduce the severity of bloating, diarrhea, constipation, and gas by increasing the amount of Lactobacillus and Bifidobacterium in the gut. The high probiotic content found in fermented foods can also boost the immune system and reduce the risk of several infections. Probiotic food can help speed up the recovery process if you're ill. Most fermented food is high in zinc, iron, and vitamin C, all great for the immune system. The broken-down nutrients in food resulting from the fermentation process can make food much easier to digest. That is why those who suffer from lactose intolerance can consume fermented food like kefir and yogurt. Antinutrients such as lectins and phytates found in grains, nuts, seeds, and legumes are broken down and destroyed during the fermentation process. Antinutrients can disrupt nutrient absorption. Research published in *Nutrients Journal* as part of a study funded by the European Regional Development Fund linked Lactobacillus rhamnosus and Lactobacillus gasseri, and other probiotic strains to reduced body fat and weight loss. Lactobacillus helveticus and Bifidobacterium longum were also linked to reduced symptoms of depression and anxiety. Fermented food may also help to lower total LDL cholesterol and blood pressure, lowering the risk of heart disease.

Choosing the Right Equipment

To prepare your vegetables for fermenting, you need to invest in a quality knife or food processor. If you are fermenting food that ferments in its own juices, such as sauerkraut, you will need a pounding tool. This will help you to compress and break apart your food in the fermenting vessel. Your choice of jars and containers will heavily depend on the type of food that is being fermented. Glass containers are easy to find and are ideal in the sense that they don't scratch easily and are free of chemicals such as BPA. Ceramic containers are typically great for large batches of fermented vegetables, as they range from 5 to 20 liters in size. Porcelain containers are generally safe, as long as you avoid pieces that are not food grade, like decorative pottery or vases. You should avoid plastic containers as they are easy to damage and scratch, easily harboring bacteria. Plastic also usually contains harmful chemicals that can impact your food. To ensure that oxygen is kept out and the gases from fermentation don't escape, you must make sure that your lid is sealed tightly. Using an airlock, a tight lid or a cloth cover are the most popular options. Airlocks are great because they reduce the risk of yeast and mold formation. Tightly sealing the lid is a good option, too. However, you may need to burp the jar daily to rid it of carbon dioxide buildup. Cloth covers may suffice, as well, but they may lead to the formation of mold and yeast.

Preparing the Vegetables

There are several ways in which you can prepare your vegetables for the fermentation process. The first is grating—you can either use your hand or a processor to grate your vegetables. Grating works well for crunchy or hard vegetables and can allow the salt to rapidly penetrate the vegetable because of the large surface area that it creates. If you grate your vegetables, it's unlikely that you'll need to add brine. You will also usually end up with a relish texture when done grating. Chopping your vegetables is also another option. Typically, the recipe that you follow should specify the size that you should chop your vegetables. However, if not specified, you can chop according to your preference. Most of the time, chopped vegetables demand salt brine. The time they take to culture will depend on their size. The third method is slicing. If you are working with firm vegetables, you can slice them thinly. If not, slice your soft vegetables into larger pieces. This will allow them to maintain their shape as they ferment. Sliced vegetables take a reasonable amount of time to ferment—not too slow and not too quick. The last method would be leaving your vegetables whole. Some vegetables such as Brussels sprouts, radishes, and green beans work best that way. Some recipes may suggest that you drown the whole vegetable in brine, while other larger vegetables may require special culture treatment.

Salt, Whey, and Starter Culture

Salt prevents bad bacteria from growing as it pulls out the moisture in the food and allows the natural bacteria that exist on the vegetable to carry out the fermentation process. This results in a slower fermentation process, ideal for cultured vegetables that need to be stored for more time. Salt also enhances the flavor of the vegetables and leaves them crunchy to bite, as it hardens the pectins in vegetables. Although salt-free ferments are more bio-diverse, they result in mushy vegetables and even mold. Various freeze-dried starter cultures can be used alone without salt. You can also use some kind of bacterial starter to speed up the fermentation process. Since whey is dairy-based, it may not be suitable for everyone. If you decide to use it, make sure that it's fresh and well-strained because it can lend its flavor to the vegetable. You can add salt to the whey for flavor and crunch.

Brine and Fermentation Recipe

1. To make the brine, you will need to follow a ratio of two tablespoons of salt to one quart of water. If the temperature is above 85 degrees, add one more spoon of salt. Stir well and set aside.

2. Prepare your vegetables in the desired method.

3. Gather your preferred spices. You can use fresh herbs, onions, and garlic.

4. Add the herbs, onions, and garlic to the bottom of your container.

5. Add mesquite leaves, black tea, horseradish, oak, or grapes to keep your vegetable crisp.

6. Place the prepared vegetable on top of the flavorings, ensuring there's 2 inches headspace between it and the top of the jar. Pour the brine over the vegetable, covering it by at least 1 inch.

7. Use an inverted plastic jar lid or a glass weight to keep your vegetable below the brine as it ferments.

8. Use your preferred lid to seal the jar tightly. Place it at 65-85 degrees Fahrenheit for around ten days (depending on your preference). The longer the time, the sourer the vegetables.

9. Burp your jar frequently if needed.

10. Once you are done, move the jar to a cold storage area.

Preventing Cross-Contamination

To prevent your cultures from cross-contamination, you should keep them at least 4 feet away from each other. Make sure that all your tools and your environment, in general, are clean. Rinse everything thoroughly to rid it of soap residue. You should always avoid antibacterial detergents and soap when fermenting and never allow your pets inside the kitchen. You should empty the space from anything that can be spilled easily and keep the ferment away from bacteria sources (garbage cans, sink drains, or soiled laundry).

Fermentation is a great way to preserve your vegetables, add other health and nutritional benefits to them, and give them a unique flavor. Fermentation is a process that has been carried out for thousands of years. Many studies are still being carried out to discover more about fermentation and its benefits. Besides aiding in digestion and boosting immunity, fermentation has also been linked to weight loss, boosting mental health, and lowering the risk of heart disease. There are several fermentation processes and vegetable preparation procedures that you can choose from—it all comes down to your preferences. You can experiment with several fermentation recipes. As long as you make sure you're using the right equipment, sealing the lid tightly, burping it when needed, and taking the necessary precautions to prevent cross-contamination.

Chapter 11: Dehydrating as a Method of Preservation

In this chapter, you will learn about a secret to food preservation that people have been using for thousands of years in times of emergency and famine: dehydrating. It turns out a dehydrated potato isn't as bad as it sounds (unless you take it personally).

Centuries ago, the only means of preserving food was to dry it, specifically to use it in winter, when it was impossible to grow fresh harvest. It's said that food dehydration originated among Middle Eastern and Oriental cultures, as it enabled ancient men and women to preserve their food sources. Egyptians used the desert heat to dry their fish and meat, and around the year 700 AD, the Aztecs began salting and sun-drying their tomatoes to conserve their freshness. Later, the Italians started to dry their tomatoes on their rooftops as well.

Dehydration is the removal of the moisture in food through evaporation to prevent microorganisms from spoiling it. Microorganisms love moist environments, so by destroying their shelter, we extend our food's shelf life.

We're all familiar with dried fruits and jerky—these are actually commercially dehydrated foods that we've come to love and consume regularly. Commercial food industries perform more efficient dehydrating processes, so the dehydrated foods that we buy can last up to ten years. Foods dehydrated at home last for shorter periods, depending on the type of food. But drying food at home is very convenient and retains the foods' nutritional value. There are many other reasons to do it at home.

DIY food dehydration:

• It can allow you to dehydrate and preserve your favorite foods safely, without harmful additives or artificial colors.

• It provides you more control over the quality and process.

• It can prove to be a lot cheaper because you'll be able to produce larger quantities any time you like.

• It is quite enjoyable, especially if you add your own touches and do it with family and friends.

Why Dehydrate Food in General?

1. Dehydrating preserves perishable foods like fresh fruits and meat, allowing you to carry them around anywhere.

2. Drying food shrinks its size while retaining the nutrients so that you can store more food in less space.

3. Dried food can reduce the time you spend cooking. You can prepare all your vegetables and legumes at once, dehydrate them, then easily rehydrate later.

Best Foods to Dehydrate

• Bananas: we're all guilty of buying more bananas than we need when we go grocery shopping, and of course, they over ripen in no time. So, it's quite convenient to peel those bananas, cut them into slices, dehydrate them, and voila! Banana chips that can last for months. Who doesn't love those?

• Apples: a nutritious fruit that, when dried, can last up to three years and can even be preserved for much longer if frozen. Dried fruits are also healthy snacks and can be added to your breakfast oatmeal.

• Beans, peas, corn, lemon: You can throw them in soups or stews, and they rehydrate immediately as you're cooking.

• Other vegetables: dried, crunchy vegetables can be dipped in sauces, grated in casseroles and salads, or even blended and used as powders (onion powder, garlic powder, etc.)

• Walla Walla onions: many people love these onions. Some even eat them like apples, but, unfortunately, they are difficult crops—their harvest lasts for only a few weeks in summer every year. Lucky for you, you can wait for the Walla Walla season, buy these tasty French onions and dehydrate them. You won't have to wait for the next season because you'll have enough supply of these sweet onions to last you the rest of the year.

• Meats: years ago, Americans decided to smoke dry meat, and, from there, jerky was invented. These thin strips of meat are very popular and delicious snacks. Dehydrating jerky at home is more economical than buying it. Finished jerky products cost about $32 per pound, while homemade jerky will cost you only $12.50 per pound. You can use any type of meat to make jerky (beef, lamb, or poultry). If you're a huge fan of jerky, here's a mouthwatering recipe!

Jerky Recipe

1. Soak the meat in vinegar, drain it, and marinate with soy sauce, salt, black pepper, and onion powder. You can add any spices you love to enhance its flavor (chili, garlic powder, or red pepper). For a sweet and sour jerky, add ¼ cup of maple syrup.

2. Place in a Ziploc bag and refrigerate for four hours or a whole day to allow the flavors to be absorbed.

3. Take the bag out and cut your meat, making sure it's uniformly sliced. Then lay the strips on the dehydrator tray and dry at 160°F for 4-6 hours. (Warning: your kitchen is going to smell amazing!)

4. Blot the meat with paper towels during the dehydration process to absorb any fat.

5. Once it reaches the level of dryness you desire, the jerky is good to go!

• Did you know that you can dehydrate your grain products? You can cook your pasta like you normally do, drain it well, spread it evenly on your tray, and dehydrate at 135°F until it's crispy and brittle. Rice can be cooked in non-fat broth before drying for an excellent taste.

• Even sauces can be easily dehydrated, spaghetti sauce, curry sauce, and salsa. They'll come in handy whenever you're too busy to make dinner. All you have to do is add some water and reheat them, and they'll rehydrate back instantly, so you can just add them to your pasta.

How to Dehydrate Food

There are various techniques for food dehydration, but regardless of the technique used, dehydration requires a few basic conditions. An adequate amount of heat to dehydrate the moisture, dry air to absorb it, and enough circulation to get rid of the vapor.

1. The Traditional Way

The traditional way of drying food would be sun drying. It's how food dehydration started and how it was done for thousands of years. Sun-drying is the easiest and cheapest way to dehydrate food, especially fruits and vegetables.

- Sun drying is ideal if you live in a hot area, but it won't work in humid or cold areas.

- All you have to do is set the food out on a tray, cover it with a cloth, expose it to direct sunlight and leave it for 3-4 days to dry.

The downside of this dehydration method is that it relies on the weather, so if temperatures drop below 85 degrees Fahrenheit, microorganisms can thrive and spoil your food.

2. Dehydrators

Dehydrators produce the best results out of most drying methods, as they have electric heating elements and air vents. Moreover, they are safe and cost-effective. A dehydrator can provide an adequate amount of heat to dehydrate the food but not enough heat to cook it (125-150 degrees F). There are a plethora of different brands and models that can give you satisfactory results. But you need to consider a few things when you're buying a dehydrator:

• Heating element: some dehydrators have the heating element on top, blowing the heat currents downwards. If the heating element is on the top or bottom, you'll have to flip the food every couple of minutes to ensure an even distribution of heat.

• Fans: your best option would be a box-type dehydrator that blows the heat currents from the back to the front and has dual fans, as it will provide a more balanced circulation inside, and there won't be a need for rotation. This makes the unit more efficient and achieves better results.

• Temperature settings: the adjustable temperatures feature is a must to dehydrate different ingredients. Many dehydrators even have timers to shut down the machine and prevent it from overheating.

- Capacity: the larger the capacity, the better. A 9-tray capacity would be ideal and will hold a lot of food. But it all depends on your preferences and how much food you'll be drying.

- Other features: If you want trays that are dishwasher safe, avoid dehydrators made of plastic and look for metal ones instead. However, they are more expensive. Some dehydrators also have transparent glass doors so that you check up on the food without opening the machine.

3. Kitchen Oven

- Ovens take much longer to dry food when compared to dehydrators because they often don't have fans. However, you can place a fan nearby.

- Your oven should go as low as 140 degrees so that your food doesn't cook instead of dry, but some ovens don't go that low.

- You can use an oven thermometer to know the exact drying temperature, as oven dials usually don't give accurate readings.

- Don't close the oven door completely. Prop it open a few inches.

4. Air Drying

Air drying takes place in enclosed spaces, unlike sun-drying. It's similar to a greenhouse, where you can control the environment during the process. You can do it in a well-ventilated attic or a dry room. Mushrooms, peppers, and herbs are all best to air-dry. But this technique may not be successful for all foods.

Storing Dehydrated Foods

The shelf life of dried food depends on storage conditions.

- •After the dehydration process, you should leave the food to cool for about forty minutes.

- •To make sure that the food is ready to be stored, check for its dryness. Fruits should feel like leather when you touch them and be slightly translucent, while vegetables should be brittle.

- •Choosing proper storage containers is essential as it prolongs shelf life, saves nutrients, prevents bugs and microbes from reaching the food, and keeps moisture out. Use metal or glass containers with tight, insect-proof, and well-insulated lids.

- •Check the food within a week or two. If there's any moisture present, you could take the food out and redry it.

- •Dried food should be kept in dark, cool, and dry places. Avoid light, as it breaks down the food and makes it lose its flavor, shortening its shelf life.

- •Fruits have double the shelf life of vegetables. Most fruit can be stored for a year, while vegetables only last for six months on average. Meat lasts up to two months only.

Dehydrating Do's and Don'ts

1. Some foods shouldn't be dehydrated, like dairy products, as the probiotics can result in food poisoning. Foods like eggs are susceptible to salmonella bacteria, and the high heat used in dehydration increases that risk. Generally, food high in fat doesn't dehydrate well, as fats don't have much moisture. Hence, they can't evaporate, leading to the food going rancid.

2. Many foods with a waxy coating, like fruits and vegetables (berries, grapes, tomatoes), should be blanched first. Blanching is plunging food from boiling water into cold water. This process

preserves color, nutrients, stops enzyme activity, slows ripening, and prevents food from spoiling or rotting quickly.

3. Keep in mind that some types of food are susceptible to oxidation, like apples and pears. You can juice any citric fruit like lemons or oranges and apply it as a pre-treatment solution. Another option would be adding a teaspoon of ascorbic acid to some water to soak the fruits in before drying them—this can help prevent most food from browning.

4. You have to use the right temperature. Too high of a temperature can result in case hardening, which dries the food's exterior and prevents the interior from drying and removing the moisture properly. In contrast, low temperatures cause the growth of bacteria and molds.

5. Don't turn up the heat when you want to speed up the drying process. It's better to slice or cut your food into smaller pieces.

6. Different foods have varying dehydration temperatures, so avoid grouping different foods together.

7. Meat and poultry need around 160 degrees Fahrenheit, fruits and vegetables need 125-135 degrees Fahrenheit, while grains and pre-cooked meats need 145 degrees Fahrenheit.

8. Don't leave the food out for too long to avoid the moisture from re-entering, and if the container is musty, it's better to throw it away or sterilize it.

9. The only vitamin that gets damaged in the dehydration process is vitamin C. So, keep vitamin C supplements on your shelf if you won't get enough from all the dried food you'll be eating.

Chapter 12: Freezing

From frozen vegetables to ready-made meals, almost any food you can think of can be found in the freezer section at the grocery store. Frozen food has been rapidly growing in popularity, offering more variety than ever since it was first introduced in the 1930s. Frozen food serves as a convenient alternative to cooking from scratch—it is time-saving, easy to cook, and can usually be prepared in multiple ways. Besides the practicality that it offers, freezing food is a great way to preserve it. Many people choose to cook and freeze large batches of food, allowing them to safely store it for long periods and reheat it whenever needed. Freezing has been used as a method to preserve food since prehistoric times. Even back then, people frequently preserved their hunt using ice and snow.

Many people believe that freezing can negatively impact the food's nutrient content. However, this is not always the case. If you blanch or submerge your fruits and vegetables in boiling water before freezing them—a method that deactivates yeasts and enzymes that may contribute to spoilage—15% to 20% of their vitamin C content can be lost. This isn't usually a problem since frozen fruits and vegetables are usually frozen right after harvesting when in peak condition. This makes them higher in nutrient content than other fresh crops that take time to be sorted, transported, and distributed. This process slowly strips them of their nutrients and vitamins. As a matter of fact, green vegetables and soft fruit can lose around 15% of their vitamin C content every day when stored at room temperature. This means that frozen and fresh fruits and vegetables end up with around the same vitamin C content.

If you freeze poultry, meat, or fish, there will be almost no mineral or vitamin loss. This is because vitamins A and D, minerals, and protein are not impacted by freezing. However, when they defrost, they lose liquids that contain mineral salts and water-soluble vitamins. If you don't recover the lost liquid, it will be completely stripped away during the cooking process.

How and Why Freezing Works

By freezing your food, you are delaying enzyme activity that leads to spoilage, as well as preventing harmful microorganisms, such as mold and yeast, from growing. This helps keep your food safe from spoilage for longer and keeps it safe for consumption. Microorganisms need the water content present in food to survive and grow. When this water turns into ice crystals upon freezing, it becomes inaccessible to them. However, you must keep in mind that some—or the majority—of the microorganisms, excluding parasites, can survive the frozen environment. This is why you must always handle your frozen food very carefully, before freezing and upon defrosting. Generally, food can last anywhere between 3 to 12 months in your freezer without spoiling or losing its quality. The specific duration depends on the type of food and the product label. If applicable, check the label and instructions for directions, special requirements, and suitable duration.

Tips on Freezing

Freezing Fruits

If you want to freeze your fruits, make sure to remove cores, stems, and other inedible parts of the fruit before washing, drying, and slicing them up. You should then place a single layer of fruit on a baking tray lined with a silicone sheet liner or compostable parchment paper. Place the fruit in the freezer for about two to three hours—or until it's completely frozen. While this step is not necessary, it is highly recommended. Freezing your fruit in a tray with a baking sheet will ensure that it won't stick together when frozen again. Ending up with a large blob of frozen fruit slices can be a pain. After it's done freezing, you can transfer the fruit slices into the container of your choice. Frozen fruit typically lasts for a long time, though it's best consumed within six to nine months.

You can either thaw the fruit or use it frozen in your recipes (if apt). Some fruits, like grapes and blueberries, are very delicious when consumed frozen. If you want to thaw your fruit, make sure to thaw only the amount that you want. The consistency of the fruit can be changed when frozen and then thawed. Freezing it for the second time may be a bad idea. You can either transfer the fruit to your fridge so that it thaws at its own pace or by running cold water over the container.

Freezing Vegetables

To freeze your vegetables, you will need to trim off the roots, stems, and other blemished areas. If you are working with vegetables that come with outer layers, like beans or peas, you should remove them. If you usually peel, core, or de-seed any of your vegetables, do so before freezing as well. Make sure that your vegetables are washed, dried, and then chopped. As mentioned above, some vegetables work best when blanched—boil some water in a pot and then submerge your vegetables in it. Keep the vegetables in long enough so that they are still firm and not cooked through. If you are working with herbs or greens, they should turn a brighter green color and wilt slightly. Scoop the vegetables out of the pot and transfer them into an ice bath. This is necessary to "shock" the vegetables, interrupting the cooking process. After your vegetables cool down, make sure to pat them dry to prevent any freezer burns. Some vegetables, like celery, bell peppers, and onions, don't need to be blanched, so make sure to find out what's best for the vegetables you are freezing. Like fruit, you should freeze your vegetables on a baking sheet before transferring them to your desired container. If properly frozen, vegetables can last around eight to ten months and should be thawed in the refrigerator or under cold water.

Freezing Fresh Meat

When freezing fresh meat, avoid freezing large amounts at once, as this helps to make the freezing and thawing processes easier. Cut and store meat in meal-sized portions to prevent it from being exposed to bacteria and make sure that it thaws easily. You should always freeze meat as fast as possible. To ensure that you are packaging your meat correctly, use a vacuum sealer or aluminum foil, plastic wrap, airtight Ziploc bags, or freezer wrap to double wrap meat cuts. Never thaw your meat at room temperature, as this makes it more vulnerable to collecting bacteria. You can thaw your meat in the refrigerator, a cold-water bath, or the microwave. Meat that has been thawed in the refrigerator can be refrozen. If you thaw it in a cold bath, make sure to change the water every thirty minutes to protect it from bacteria. Microwaving is ideal if you are under a time crunch. However, it can change the meat's consistency and texture.

Freezing Seafood

When it comes to seafood, freezing is not a one-size-fits-all process. For every 500 gm of lobster, cook in salted water for eight minutes, drain, and cool. Place in a plastic bag, cover it with brine, seal, and freeze. Scallops need to be rinsed under cold water and free of sand and their shell. Drain, place in a plastic bag, seal, and freeze as well. For clams and mussels, rinse them to get rid of sand, then steam them over medium heat so that the shells open slightly. Shuck them, place them in a plastic bag, cover them in brine or the strained cooking liquid, seal, and freeze. Oysters must be shucked, rinsed, and reserved in liquid. You should then pack them and their liquid in a plastic bag, then seal and freeze them. Finfish should generally be gutted and rinsed under cold water. Wrap small fish in baking sheets, then freeze them—large fish should be frozen on a baking sheet, dipped in ice-cold water, and returned to the freezer. Repeat multiple times, wrap well, and freeze. It's recommended that you cover steaks and fillets with brine solution before freezing in a plastic bag or container. The recommended freezing time varies based on the type of seafood, though seafood usually lasts between three to six months when handled correctly. You can thaw seafood in the refrigerator or under cold water—don't unfreeze until you're ready to cook.

Freezing Eggs and Dairy

You can freeze your eggs for up to one year. To freeze the eggs, you should remove the shells and scramble them well. Pour the mixed eggs into an ice cube tray, wrap them well using saran wrap, and freeze. After they freeze, you can either leave them in the tray or transfer them into another container. Eggs need to be thawed in a refrigerator. Milk can be frozen. However, keep in mind that it may separate when thawed. Make sure to remove 1 ½ cups per gallon when freezing the milk. This will guarantee that the milk doesn't burst the container or overflow when it expands. Frozen milk can last up to one month and can be thawed in the refrigerator. Make sure to shake it well before you use it.

Freezing Soups, Broths, and Stews

You can use rectangular plastic containers to freeze your soups, broths, and stews. Line your container with saran wrap and make sure that it hangs over the sides. Fill it up with a pre-measured amount of soup/ broth/ stew, and then place the container on a rimmed sheet pan. Freeze them flat, flip out the container, and wrap the saran wrap around your frozen soup, broth, or stew. You can place the wrapped product in a Ziploc bag. To thaw, heat it over low or medium heat until it thaws completely.

Which Containers to Use

Ideally, you should use re-sealable freezer bags, glass storage containers that come with airtight lids, or plastic deli containers to freeze your food. Some wide jars are also suitable for freezing. However, in all cases, make sure you read the label to ensure they're freezer friendly. You should also leave one-inch headspace between the lid and whatever you are freezing. Keep in mind that when you freeze liquids, they expand. In which case, glass containers, if not freezer-safe, may become hazardous and shatter. This is why random glass jars may not be the best option if you are freezing milk or sauces. Besides unreliable glass, their seals may not be airtight. As explained above, the best packaging method and most suitable type of container depends on the food you are freezing. If you choose the correct containers and follow packaging and thawing procedures, you will avoid off-flavors. You should also stick to the recommended freezing period to maintain flavor, texture, and consistency.

Freezer Tips

Your freezer should always be kept at -18°C or below the suggested temperature. While refrigerators may not function when they are full, deep freezers function better when packed tightly. Don't confuse this with overloading, however. Overloading your freezer can block the vents, reducing their lifespan. It may also damage your food due to blockage of air circulation, catalyzing bacterial growth. Over packing your freezer can also impact its energy efficiency. To avoid freezer burns, messes, or spills, ensure that everything is packed in suitable and freezer-safe containers. You should always avoid placing hot food in the freezer. This is why you should run your vegetables under cold water or place them in ice baths before freezing. Placing hot food in the freezer will raise the freezer temperature. Not only will this negatively impact your freezer, but it may also ruin other food, as it could end up melting or producing water. When the food thaws and refreezes incorrectly, it can develop an off-taste. This could also mess with its consistency and even attract bacteria. You should always make sure that your food is completely thawed before you cook it. Don't refreeze your food after it has been cooked.

To prevent your freezer from over packing, you should frequently empty it and rid it of all expired products. Store-bought frozen food often comes in bulky, useless boxes. Tossing these boxes out and replacing them with Ziplocs or freezer bags can help you to save space. Another excellent freezer organization tip is storing everything vertically instead of laying it flat. You will be surprised by how much space you will gain. Besides, keeping things vertically will allow you to have a better view of everything. You can also use paper clips strategically to hang things from the shelves and store the larger items at the bottom of your freezer.

What to Do When a Freezer Stops Working

If the energy goes out or your freezer stops working for whatever reason, there are a few things that you should do to ensure that your food doesn't go bad. If you know that the issue can be fixed within a day, then make sure to keep your freezer shut. Every time you open your freezer, cold air will escape, and warm air will flow in. If your freezer doesn't have a thermostat, you should get a compatible one. This way, when the freezer stops working, you will determine the status of your food. As long as your freezer doesn't go above 4°C, it should be safe. However, keep in mind that not all food is the same and that this is just a general rule. If your freezer isn't fixed within a day, you should take some alternate protective measures. One of the easiest options is to ask a friend, neighbor, or family member if they have any room in their freezer. Make sure to transport your food in an icebox. You can also use dry ice to keep your freezer temporarily cold. However, be very careful when handling dry ice to avoid chemical burns. Use gloves to wrap the dry ice in newspapers or towels before placing it in the freezer. You can also use insulated coolers or an icebox to keep your food at home. Make sure that the temperature doesn't rise above 4°C.

Freezing has been used for centuries as a food preservation method. It is a great way to ensure that you have ready-made, safe food whenever you don't feel like cooking. It is also a great way to preserve fruit and vegetables for later consumption. Many people freeze off-season fruits to have access to them all year round. For instance, they may freeze mangoes during summer to make fresh mango juice during winter when unavailable at the markets. Freezing comes with endless advantages. However, to ensure that your food remains bacteria-free, doesn't develop an off-taste, or change its texture and consistency, you must keep several things in mind. Based on the food you are handling, follow through with the suitable preparation, packing, freezing, and thawing procedures. You should also make sure to use freezer-safe containers suitable for the type of frozen food you will put inside them to steer clear of potential hazards.

Conclusion

Around 1.3 billion tons of food are wasted around the world each year. Many people buy food that stays forgotten at the back of the pantry. Others may buy an excessive amount of food because they overestimate their needs. Regardless of the reason, food spoilage and waste remains among the most popular problems our world faces. Another issue we encounter is seasonal dietary restrictions or limitations. Many fruits and vegetables are seasonal. They are only available or more abundant during specific times of the year. This could be a major problem for some people if their diet heavily depends on specific seasonal foods. For others, craving a food item when it's less abundant can be a minor inconvenience. Whichever problems you're facing, they can be fixed through canning and preserving.

Now that you have read this book, you probably know everything there is to know about canning and preserving. It's time to get your hands on a canner, mason jars, a food processor, a dehydrator, a juicer, and any other supplies you may need. Make sure to select the right supplies for the job, though! For instance, if you're freezing your food, make sure to invest in freezer-safe containers, or get the right jars and lids if you plan to ferment. Before you jump right in, make sure that you conduct the right preservation technique for the type of food, your needs, and desired outcome. Knowing exactly what you expect from the process will help you to figure out which technique is the most suitable. Following through with the right procedure will help to guarantee a successful preservation experience. When it's time to use your preserved food, you should also refer to the instructions when needed.

Understanding how you can preserve your food will surely change your life. As you put all this information into practice, you will find that the time you spend in the kitchen drastically decreases. You will also notice that less food is being wasted, your diet is becoming more diverse, and your grocery shopping bills are decreasing. When you preserve your food at home, you control everything that occurs throughout the process. Being aware of the equipment used, the cleanliness of the supplies, the general environmental conditions, and, of course, avoiding the additives that suppliers use will keep your mind at ease. It is a great way to ensure that your family's health remains a top priority, while still benefiting from everything that preservation has to offer. While processing and implementing all this information in real-time may feel very overwhelming at first, preserving your food will become a habitual and regular kitchen routine.

Whether you want to save your chicken broth for later or make batches of semi-cooked food to prepare whenever you need it, you will automatically put all your knowledge into practice. Besides, if you ever feel stuck, you will always have this friendly handbook to turn to. Filled with simple, delicious recipes and step-by-step directions, you will eventually master the art of making delicious homemade jams and nutritious preserved goods. Whether you are into pickling or fermenting, you will find a way to give your food a unique flavor and please your family's taste buds. Knowing how to preserve your own food may just be the skill that you never knew you'd need until you started it.

Here's another book by
Dion Rosser that you might like

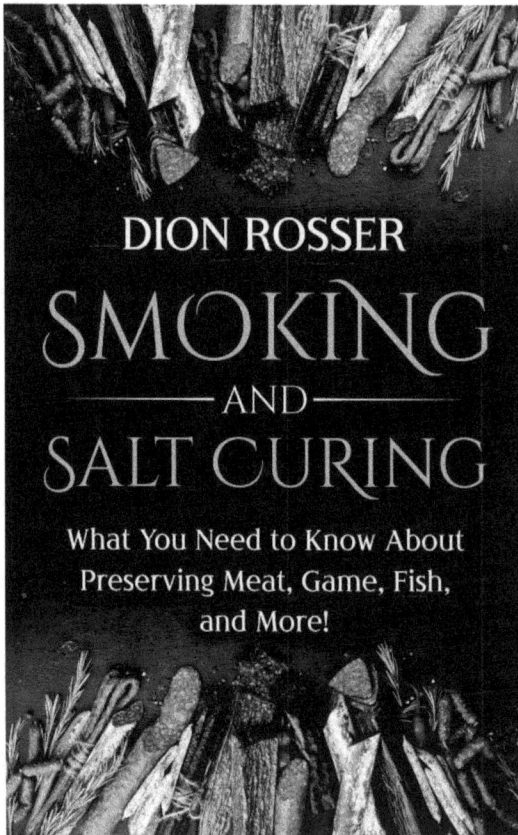

DION ROSSER

SMOKING
—AND—
SALT CURING

What You Need to Know About
Preserving Meat, Game, Fish,
and More!

References

Ashiya. (2011, May 17). *What Is the Importance of Food Preservation?*
https://www.preservearticles.com/articles/what-is-the-importance-of-food-
preservation/5187

Approved Canning Methods: Types of Canners. (n.d.)
https://extension.psu.edu/approved-canning-methods-types-of-canners

Ewald, J. (2014, August 8). *What is Canning, and What Are the Benefits?* 21, 2021.
https://lifeandhealth.org/lifestyle/what-is-canning-and-what-are-the-
benefits/172324.html

Guerrero-Legarreta, I. (2004). *Canning. Encyclopedia of Meat Sciences.* (pp. 139–
144). Elsevier.

National center for home food preservation. (n.d.)
https://nchfp.uga.edu/publications/nchfp/factsheets/food_pres_hist.html

Neverman, L. (2021, January 16). *Home Food Preservation – Ten Ways to
Preserve Food at Home.* https://commonsensehome.com/home-food-preservation/

Athearn, Kevin & Simonne, Amarat & Ahn, Soohyoun. (2018). *Budget Template
for Home Canning.* EDIS. 2018.

Christensen, E. (2010, July 30). *Canning Basics: What's the Deal with Pectin?*
https://www.thekitchn.com/canning-basics-whats-the-deal-123192

Cook's Info. (2005, May 19). *Canning Funnels. Cook's Info.*
https://www.cooksinfo.com/canning-funnels

Food in Jars. (2010, July 28). *Canning 101: Why You Should Bubble Your Jars.*
https://foodinjars.com/blog/canning-101-why-you-should-bubble-your-jars/

Healthy Canning. (n.d.). *Pickling Spice.* https://www.healthycanning.com/pickling-spice/

Healthy Canning. (n.d.). *Sugar's Role in Home Canning.* https://www.healthycanning.com/sugars-role-in-home-canning/

Healthy Canning. (n.d.). *Vinegar in Canning Water.* https://www.healthycanning.com/vinegar-in-canning-water/

Helseth, R. (2021, 18). *Wide Mouth Vs. Regular Mouth Mason Jars – The Same, But Different.* https://masonjarlifestyle.com/wide-mouth-vs-regular-mouth-mason-jars-the-same-but-different/

Joachim, D., & Schloss, A. (n.d.). *The Science of Pectin. Fine Cooking.* https://www.finecooking.com/article/the-science-of-pectin

Loe, T. (n.d.). *Pickling Salt vs. Other Salts in Canning.* https://livinghomegrown.com/pickling-salt/

Michelle. (2012, September 12). *Canning Equipment 101: The Tools You Need to Start Canning.* https://rosybluhome.com/canning-equipment-101-the-tools-you-need-to-start-canning/

Sarah. (2021, 16). *Canning Supplies and Preserving Equipment List. Sustainable Thoughts.* https://www.sustainablecooks.com/canning-supplies/

How to Select the Best Fruit for Jam/Jellies. (n.d.) https://www.vigopresses.co.uk/AdditionalDepartments/Right-hand-panel/Vigo-Presses-Blog/How-to-Select-the-Best-Fruit-for-JamJellies

Lebert, A. (2017). *Fermented Meat Products. Current Developments in Biotechnology and Bioengineering* (pp. 25–43). Elsevier.

Sharon. (2020a, 14). *Canning Chicken: It's Great for Homemade Soup or Casserole Recipes.* https://www.simplycanning.com/canning-chicken/

Sharon. (2020b, 19). *Canning Venison – Raw Packed, Cubed, or Strips. It's So Easy!* https://www.simplycanning.com/canning-venison/

Sharon. (2020, May 5). *Ground Venison.* https://www.simplycanning.com/canning-venison-ground/

The Meadow. (n.d.). *Making Fermented Sausage.* https://themeadow.com/pages/making-fermented-sausage

Utah State University. (n.d.). *Canning Meats, Poultry, and Seafood.* https://extension.usu.edu/preserve-the-harvest/research/canning-meats-poultry-seafood

Winger, J. (2020, June 29). *How to Can Food with No Special Equipment.* https://www.theprairiehomestead.com/2020/06/canning-no-equipment.html

Beyond Sauerkraut: A Brief History of Fermented Foods. (2014, March 3). https://www.lhf.org/2014/03/beyond-sauerkraut-a-brief-history-of-fermented-foods/

Foodelicious. (n.d.). *Catherine's Pickled Blueberries.* https://www.allrecipes.com/recipe/206655/catherines-pickled-blueberries/

Imatome-Yun, N. (2015, January 24). *Save Your Pickle Juice for Pickletinis, Hangovers, & More.* https://food-hacks.wonderhowto.com/how-to/save-your-pickle-juice-for-pickletinis-hangovers-more-0159741/

Marisa. (2013, 10). *Preserving Spring: Spicy Pickled Asparagus.* https://simplebites.net/preserving-spring-spicy-pickled-asparagus-recipe/

Pickled Fruit. (2020, September 2). https://wavesinthekitchen.com/pickled-fruit/

Pruitt, S. (2015, May 21). *The Juicy 4,000-Year History of Pickles.* https://www.history.com/news/pickles-history-timeline

Sevier, J. (2017, August 11). *How to Can Pickles, Step by Step.* https://www.epicurious.com/expert-advice/how-to-can-pickles-weekend-warrior-article

Lang, A., BSc, & MBA. (2019, December 10). *Jam vs. Jelly: What's the Difference?* https://www.healthline.com/nutrition/jelly-vs-jam

Martens, M. F. A. (2020, June 29). *A Guide to Homemade Jams and Jellies.* https://asweatlife.com/2020/06/how-to-make-your-own-jam/

Beaty, V. (2018, June 26). *Thirty Homemade Fruit Jams and Jellies You Definitely Want to Make this Summer.* https://www.diyncrafts.com/41444/food/30-homemade-fruit-jams-and-jellies-you-definitely-want-to-make-this-summer

Chihak, S. (2018, July 26). *How to Use a Pressure Canner to Preserve Your Veggies, Meats, and More.* https://www.bhg.com/recipes/how-to/preserving-canning/pressure-canning-basics/

Meredith, L. (n.d.). *Boiling Water Bath vs. Pressure Canning.* https://www.thespruceeats.com/boiling-water-bath-versus-pressure-canning-1327438

Old Farmer's Almanac. (n.d.) *Pressure Canning: Beginner's Guide and Recipes.* https://www.almanac.com/pressure-canning-guide

Sharon. (2020, 29). *Pressure Canning, Learn How to Use Your Pressure Canner.* https://www.simplycanning.com/pressure-canning/

Autumn. (2018, October 25). *Pressure Canning Fish: A Basic Recipe.* 21, 2021. https://atraditionallife.com/pressure-canning-fish-basic-recipe/

Canning Fish in Quart Jars. (n.d.). 21, 2021. https://nchfp.uga.edu/how/can_05/alaska_can_fish_qtjars.pdf

---

How to Can Fresh Fish for Beginners. (2021, February 17). 21, 2021.
https://www.anoffgridlife.com/how-to-can-fresh-fish-for-beginners/

Williams, T. (2016, 4). *How to Can Fish (Salmon, Tuna, and More!)*
http://themasonjarsuite.squarespace.com/videos/howtocanfish

Canning Meats & Poultry. (2020, October 30).
https://www.clemson.edu/extension/food/canning/canning-tips/51canning-meats-
poultry.html

Tayse, R. (2014, July 3). *Nine Need-to-Know Tips for Canning Meat.*
https://www.hobbyfarms.com/9-need-to-know-tips-for-canning-meat-2/

Thomas, C. (2020, January 30). *Step by Step Tutorial for Canning Meat* (Raw Pack
Method). https://homesteadingfamily.com/step-by-step-tutorial-for-canning-meat-
raw-pack-method/

Winger, J. (2015, January 27). *Canning Meat – Tutorial.*
https://www.theprairiehomestead.com/2015/01/canning-meat.html

Winger, J. (2020, March 12). *Canning Chicken (How to Do it Safely).*
https://www.theprairiehomestead.com/2020/03/canning-chicken.html

Cloudflare. (n.d.). *What is Fermentation – Masterclass.*
https://www.masterclass.com/articles/what-is-fermentation-learn-about-the-3-
different-types-of-fermentation-and-6-tips-for-homemade-fermentation#what-are-the-
different-stages-of-the-fermentation-process

Coyle, D. C. (2020, August 20). *What Is Fermentation? The Lowdown on
Fermented Foods.* https://www.healthline.com/nutrition/fermentation#safety

Cultures for Health. (2016a, May 16). *How To Prepare Your Vegetables for
Fermentation.* https://www.culturesforhealth.com/learn/natural-fermentation/how-to-
prepare-vegetables-fermentation/

Cultures for Health. (2016b, June 24). *Cross-Contamination: Keeping Your
Cultures Safe from Each Other.*
https://www.culturesforhealth.com/learn/general/cross-contamination-keeping-
cultures-safe/

Cultures for Health. (2017, August 18). *Fermentation Supplies.*
https://www.culturesforhealth.com/learn/natural-fermentation/fermentation-
equipment-choosing-the-right-supplies/

Cultures for Health. (2018, June 14). *Salt vs. Whey vs. Starter Cultures for
Fermenting Vegetables, Fruits & Condiments.*
https://www.culturesforhealth.com/learn/natural-fermentation/salt-vs-whey-vs-starter-
cultures/

Cultures for Health. (2020, November 9). *How To Ferment Vegetables: Everything You Need to Know*. https://www.culturesforhealth.com/learn/natural-fermentation/how-to-ferment-vegetables/

S. (2020, June 19). *Make Old-Fashioned Brine Fermented Pickles Like Your Great Grandmother*. https://simplebites.net/make-old-fashioned-brine-fermented-pickles-like-your-great-grandmother/

W. (2018, March 27). *Beyond Sauerkraut: A Brief History of Fermented Foods*. https://www.lhf.org/2014/03/beyond-sauerkraut-a-brief-history-of-fermented-foods/

Drying Food at Home. (n.d.) https://extension.umn.edu/preserving-and-preparing/drying-food

Farm, J. (2020, July 10). *Three Easy Dehydrator Jerky Recipes for Summer Hikes and Car Trips*. https://joybileefarm.com/dehydrator-jerky-recipes/

Fresh Off the Grid. (2020, May 21). *The Ultimate Guide to Dehydrating Food*. https://www.freshoffthegrid.com/dehydrating-food/

Reed, L. (2020). *Dehydrating food: The Beginner's Guide to Dehydrating Vegetables, Fruits, Meat, and Other Foods at Home with Easy Recipes*. Tonazzi Company.

Washington Post. (2020, March 22). *How to Freeze Fresh Vegetables*. https://www.washingtonpost.com/news/voraciously/wp/2020/03/22/how-to-freeze-fresh-vegetables-while-preserving-their-best-qualities/

A. (2021, January 4). *Why You Should Never Overload Your Refrigerator or Freezer*. https://allareaappliancellc.com/why-you-should-never-overload-your-refrigerator-or-freezer/

Crigger, D. (2021, February 2). *Brilliant Freezer Organization Tips You Need*. https://www.onecrazyhouse.com/17-ways-organize-freezer-probably-arent/

Fin, L. (2019, August 20). *Best Way to Efficiently Freezer-Store Soups, Broths and Stews*. https://food52.com/recipes/16882-best-way-to-efficiently-freezer-store-soups-broths-and-stews

How Does Freezing Preserve Food and Maintain Quality? (n.d.). https://www.eufic.org/en/food-safety/article/chilling-out-freezing-foods-for-quality-and-safety

Jennings, M. (2019, February 26). *How to Freeze Milk and Eggs*. https://www.stockpilingmoms.com/frugal-friday-how-to-freeze-milk-and-eggs.

Name, T. (2020, October 23). *How to Protect Your Food When Your Freezer Stops Working*. https://www.capitalcityapplianceservice.com/blog/how-to-protect-your-food-when-your-freezer-stops-working/

R. (2020, January 9). *Tips for Freezing: A Guide to Proper Meat Storage.* https://www.beststopinscott.com/freezing-meat/

The Right Way to Freeze Fresh Fruits to Enjoy All Year Long. (2020, 14). https://www.stasherbag.com/blogs/stasher-life/the-right-way-to-freeze-your-summer-fruits.

Tips on Freezing Seafood. (2018, July 13). https://www.tourismpei.com/tips-on-freezing-seafood

Taraneh. (2021, March 1). *Best Canning and Food Preservation Equipment for Preserving Local Bounty.* https://www.kitsilano.ca/2016/10/11/best-canning-and-food-preservation-equipment

www.ingramcontent.com/pod-product-compliance
Lightning Source LLC
Chambersburg PA
CBHW071900090426
42811CB00004B/680